The Social History of Poverty:
The Urban Experience

Francesco Cordasco
EDITOR

IN DARKEST ENGLAND

AND

THE WAY OUT

BY

William Booth

With a New Foreword by

FRANCESCO CORDASCO

GARRETT PRESS, INC.
New York, 1970

SBN 512-00752-7
Library of Congress Catalog Card
Number 79-106733

The text of this book is a photographic reprint of the first edition,
published in London by the International Headquarters of the Salvation Army in 1890.
Reproduced from a copy in the Garrett Press Collection.

First Garrett Press Edition Published 1970

Manufactured in the United States of America

GARRETT PRESS, INC.
Publishers

250 West 54th Street, New York, N.Y. 10019

FOREWORD

In the history of the urban poor, the Salvation Army and the work of William [General] Booth have a special place. Organized in 1878 by William Booth, the Salvation Army began as a protest against the intellectualism of all religious sects, particularly Methodists and Quakers and the proliferating democratic groups within the Methodist communion. Like other sects the Salvation Army had set out to evangelize the urban poor, and its early attempts to preach emotional religion to the poor (largely church-unaffiliated)[1] in East London were no more successful than any other evangelical efforts.

That the Salvation Army should have become a dynamic agency of social reform was due as much to the failure of "fashionable religion" as to the "Holy War" upon which William Booth and his remarkable wife, Catherine, embarked. The Salvation Army was organized because the Booths were convinced that the poor could be made Christians only by the fellow poor who had attained the conviction of salvation: in Catherine Booth's words "by people of their own class, who would go after them in their own resorts, who would speak to them in a language

---

[1] "England tended to think of itself as a religious nation. Until the Reform Act of 1867 the number of Englishmen who had some religious affiliation was greater than the number of voters. But after 1867, and still more after 1884, the number of voters far exceeded the number of active members of Christian churches. Hence the growth of democracy compelled the English to realize. . . that English piety, especially that of the Anglican Church, might after all be another manifestation of the class system, a somewhat superficial phenomenon, the appurtenance of a few." Helen M. Lynd, *England in the Eighteen Eighties: Toward a Social Base of Freedom* (London: Oxford University Press, 1945) p. 339.

they understood, and reach them by means suited to their own tastes."[2]

William Booth (1829-1912) had wide experience in Methodist sects. He began as a Wesleyan lay preacher in Nottingham, served with the Wesleyan Reformers, and in 1858 became a minister in the Methodist New Connexion; and his wife had been a Wesleyan until expelled. As Professor K.S. Inglis observes: "Much in the history of the Salvation Army becomes intelligible only when it is seen as the creation of dissatisfied Methodists."[3] Since the Booths encountered continuing difficulties in the Methodist Connexion, they turned to other means to implement their social gospel, and it may well be true that William Booth could never have worked for long in an organization devised by others; when they launched a crusade in White-Chapel in 1865 (the "Christian Mission to the Heathen of Our Own Country"), it was out of the expansion of this crusade that the Salvation Army was to emerge, and the principles of "Salvationism" were clearly delineated in Booth's early writings.[4]

A staggering literature is devoted to the history of the Salvation Army and to the life of William Booth (the two are inextricable);[5] and if the doctrines of the Salvation

[2] F. de L. Booth-Tucker, *Catherine Booth: The Mother of the Salvation Army* (London: Sonnenschein, 1892), vol. II, p. 234.

[3] K. S. Inglis, *Churches and the Working Classes in Victorian England* (London: Routledge and Kegan Paul, 1963), p. 176.

[4] See particularly, William Booth, *How to Reach the Masses with the Gospel* (London, 1872).

[5] See the *Salvation Army Yearbooks,* generally; also, William Booth, *Orders and Regulations for Field Officers* (London, 1886); Bramwell Booth, *Echoes and Memories* (London: Hodder & Stoughton, 1925), and *These Fifty Years* (London: Cassell, 1929); St. John Irvine, *God's Soldier: General William Booth,* 2 vols. (New York: Macmillan, 1925); Harold Begpie, *Life of William Booth: Founder of the Salvation Army,* 2 vols. (London: Macmillan, 1920). The best short account of the Salvation Army set into historical context is in K. S. Inglis, *op. cit.,* pp.175-214.

Army were simply those of evangelical revival, they nonetheless had a tremendous impact on the lives of the poor; and they had an even greater impact on the responses of the churches to poverty and its myriad social problems.[6] In many ways the influence which the Salvation Army exercised cannot be easily assessed: in 1883, its journal, the *War Cry* had a circulation of 350,000 copies a week in its English edition alone, and the message and the activities of the Army were becoming worldwide.

Despite the initial indifference to temporal circumstances, and Booth's insistence that the world needed salvation alone ("You don't need to mix up any other ingredients with the heavenly remedy," he had declared in *Salvation Soldiery*), the Army had to turn its attention to the stark spectre of poverty since it was clear that poverty itself was the greatest impediment to salvation; and the Army's plan for salvation became the "Social Christianity" which William Booth formulated in *In Darkest England and the Way Out* (1890).[7] For Booth, salvation remained the only objective; the social scheme ennunciated in *In Darkest England* only altered his strategy of salvation, and in this he was very clear:

[6] The Archbishop of Canterbury, A.C. Tait, remarked that since there was "a vast mass of persons who could not be reached by the more regular administration of the Church, it was not unlikely that much good might eventually result from the more irregular action of the Salvationists." See K. S. Inglis, *op. cit.*, pp. 188 *ff.* for other reactions; and also, Warren Sylvester Smith, *The London Heretics, 1870-1914* (New York: Dodd, Mead, 1968).

[7] For an excellent appraisal of the work of the Salvation Army before 1890, and the new direction taken after 1890 and the publication of *In Darkest England,* see Robert Archey Woods, *English Social Movements,* 2nd ed. (New York: Scribner, 1895). Issued originally in London under the imprint of International Headquarters of the Salvation Army, *In Darkest England* was also published in New York (Funk & Wagnalls, 1890) and in Chicago (C.H. Sergel, 1890). The title was a paraphrase of Henry Morton Stanley's *In Darkest Africa or The Rescue of Emin* (1890) which had attracted considerable attention.

"Now, I propose to go straight for these sinking
classes, and in doing so shall continue to aim at the
heart. I still prophesy the uttermost disappointment
unless that citadel is reached. In proposing to add one
more to the methods I have already put into operation
to this end, do not let it be supposed that I am the less
dependent upon the old plans, or that I seek anything
short of the old conquest. If we help the man it is in
order that we may change him. The builder who should
elaborate his design and erect his house and risk his
reputation without burning his bricks would be pronounced a failure and a fool. Perfection of architectural
beauty, unlimited expenditure of capital, unfailing
watchfulness of his labourers, would avail him nothing if
the bricks were merely unkilned clay. Let him kindle a
fire. And so here I see the folly of hoping to accomplish
anything abiding, either in the circumstances or the
morals of these hopeless classes, except there be a
change effected in the whole man as well as in his
surroundings. To this everything I hope to attempt will
tend. In many cases I shall succeed, in some I shall fail;
but even in failing of this my ultimate design, I shall at
least benefit the bodies, if not the souls, of men; and if I
do not save the fathers, I shall make a better chance for
the children.

It will be seen, therefore, that in this or in any other
development that may follow, I have no intention to
depart in the smallest degree from the main principles on
which I have acted in the past. My only hope for the
permanent deliverance of mankind from misery, either
in this world or the next, is the regeneration of remaking
of the individual by the power of the Holy Ghost
through Jesus Christ. But in providing for the relief of
temporal misery I reckon that I am only making it easy
where it is now difficult, and possible where it is now all
but impossible, for men and women to find their way to
the Cross of Our Lord Jesus Christ." *(Preface)*

The success of the book was immediate (by 1891 it
had sold over 200,000 copies), and it inspired considerable

financial support for the work of the Army.[8] It sought no reconstruction of the social system; instead it sought to rescue the "submerged tenth" (some three million of the population, Booth estimated) who were an underclass: unemployed, homeless, vicious, criminal, with hoards of children.

> "Darkest England may be described as consisting broadly of three circles, one within the other. The outer and widest circle is inhabited by the starving and the homeless, but honest, Poor. The second by those who live by Vice; and the third and innermost region at the centre is peopled by those who exist by Crime. The whole of the three circles is sodden with Drink. Darkest England has many more public-houses than the Forest of the Aruwimi has rivers, of which Mr. Stanley sometimes had to cross three in half-an-hour.
>
> The borders of this great lost land are not sharply defined. They are continually expanding or contracting. Whenever there is a period of depression in trade, they stretch; when prosperity returns, they contract. So far as individuals are concerned, there are none among the hundreds of thousands who live upon the outskirts of the dark forest who can truly say that they or their children are secure from being hopelessly entangled in its labyrinth. The death of the bread-winner, a long illness, a failure in the City, or any one of a thousand other causes which might be named, will bring within the first circle those who at present imagine themselves free from all danger of actual want. The death-rate in Darkest England is high. Death is the great gaol-deliverer of the captives. But the dead are hardly in the grave before their places are taken by others. Some escape, but the majority, their health sapped by their surroundings, become weaker and weaker, until at last they fall by the

[8] For contemporary reviews of the book, see *Westminister Review* (April 1891); *Contemporary Review* (December 1890); *Church Quarterly Review* (April 1891) as samples of the huge literature which the book provoked. For the influence of Frank Smith, a Commissioner in the Army, and the role of W.T. Stead in the writing of the book, see K.S. Inglis, *op. cit.*, pp. 199 *ff.*

way, perishing without hope at the very doors of the
palatial mansions which, maybe, some of them helped to
build." (p. 24)

In essence, the plan proposed was simple: to bring the
unemployed into "self-helping and self-sustaining communities, each being a kind of co-operative society or
patriarchal family, governed and disciplined on the principles which have already proved so effective in the
Salvation Army." The plan included a City Colony, a Farm
Colony, and an Overseas Colony: in short, a centralized
society with Booth's organization at the center providing
direction and control. That the plan had as its only
objective salvation was nowhere more vividly evident than
in the lithographed chart ("Salvation Army Social
Campaign") tipped in as a grim frontispiece to *In Darkest
England;* and the prose of the book is demoniacal,
tendentious, exhortatory, and heuristic. Its plan for social
reform is the scaffolding for an elaborate sermon addressed
to a sinful and corrupt world.

"But this book is no mere lamentation of despair. For
Darkest England, as for Darkest Africa, there is a light
beyond. I think I see my way out, a way by which these
wretched ones may escape from the gloom of their
miserable existence into a higher and happier life. Long
wandering in the Forest of the Shadow of Death at our
doors, has familiarized me with its horrors; but while the
realisation is a vigorous spur to action it has never been
so oppressive as to extinguish hope. Mr. Stanley never
succumbed to the terrors which oppressed his followers.
He had lived in a larger life, and knew that the forest,
though long, was not interminable. Every step forward
brought him nearer his destined goal, nearer to the light
of the sun, the clear sky, and the rolling uplands of the
grazing land. Therefore he did not despair. The
Equatorial Forest was, after all, a mere corner of one
quarter of the world. In the knowledge of the light
outside, in the confidence begotten by past experience

of successful endeavour, he pressed forward; and when
the 160 days' struggle was over, he and his men came
out into a pleasant place where the land smiled with
peace and plenty, and their hardships and hunger were
forgotten in the joy of a great deliverance.
So I venture to believe it will be with us. But the end
is not yet. We are still in the depths of the depressing
gloom. It is in no spirit of light-heartedness that this
book is sent forth into the world. The magnitude of the
evils and the difficulty of dealing with them are
immense." (p. 15)

In the literature of late 19th century England which
addressed itself to poverty and its amelioration, *In Darkest
England* has a unique place. It is, to be sure, part of the
great social context in which Professor Inglis describes the
role of the churches as they responded to the poor. It is
not unlike much of the effort which went into Toynbee
Hall, and the "settlement" experiments in religious and
social action, and yet the plan it proposes is a striking
departure from anything which the "settlement" experiments attempted. It was unlike any program which had an
anchorage in any church; and in no way can it be
compared with or related to the programs and efforts
devised by secular reformers such as Charles Booth
(1840-1914) whose *Life and Labours of the People in
London* undoubtedly influenced William Booth as its early
volumes were published.[9] For the contemporary reader, *In
Darkest England* has a striking relevancy to much of the
effort which has been currently mounted against urban
poverty: its rhetoric and its ideology (stripped of its
"salvationist" elements) are not unlike our impassioned
entreaties against poverty and injustice; and its earnestness
and optimism are not unlike the evangelical war on

[9] For Charles Booth, see Harold W. Pfautz, ed., *Charles Booth on the City:
Physical Pattern and Social Structure* (Chicago: University of Chicago Press,
1967).

poverty which Johnsonian America launched in the
1960's. Above all, a reading of *In Darkest England* is both
a sad and poignant experience: it is at once and for all a
recognition that ideology (whatever its animus) is inadequate to social challenges and in our complicated and
uninnocent world truly as unavailing as it proved in the
more hopeful and innocent contexts of Boothian England.
In Professor Inglis' laconic and correct assessment:

> "William Booth was the most spectacular example of a
> Christian converted into a social reformer by what
> appeared to him the dictates of strategy. Others before
> him, however, had concluded that the social environ-
> ment of the poorest made them inaccessible to the
> gospel; and in the few years before Booth's conversion a
> number of Christians were thinking like the writer in the
> *Methodist Times* who said in 1886; 'the duty of the
> Evangelist is not simply to preach the Gospel, but if the
> condition of his hearer is unfavourable for his reception,
> it becomes his duty also to improve those conditions....'
> *In Darkest England* dramatized this notion, and helped
> to make it a commonplace after 1890 among Christians
> of all persuasions. Prudence like Booth's would not
> make a Christian an ally of the Social Democratic
> Federation, the Independent Labour Party or even the
> Fabian Society. Its object was simply the lowest stratum
> of the working classes, the people whose circumstances
> were wretched and corrupting — in Booth's word, the
> submerged. Booth and others believed that the sub-
> merged could be brought into the range of Christian
> missionaries without any other change in the social
> structure. For this reason socialists had a point when
> they argued that Booth was offering merely a gigantic
> engine for old-fashioned relief. In Booth and in others,
> the view that social reform was strictly necessary to
> evangelism could sound remarkably like an appeal for
> more and more charity."[10]

Francesco Cordasco
Montclair State College

[10] K.S. Inglis, *op. cit.*, pp. 259-260.

IN DARKEST ENGLAND AND THE WAY OUT.

PRINTED BY
MᶜCORQUODALE & CO., LIMITED,
CARDINGTON STREET, EUSTON,
LONDON, N.W.

IN DARKEST ENGLAND

AND

THE WAY OUT.

BY

GENERAL BOOTH.

London:
INTERNATIONAL HEADQUARTERS OF THE SALVATION ARMY,
101, QUEEN VICTORIA STREET, E.C.

New York:	*Melbourne:*	*Toronto:*
111, READE ST.	185, LITTLE COLLINS ST.	ALBERT ST.

TO THE MEMORY

<small>OF THE</small>

COMPANION, COUNSELLOR, AND COMRADE

OF NEARLY 40 YEARS,

THE SHARER OF MY EVERY AMBITION

<small>FOR</small>

THE WELFARE OF MANKIND,

MY

LOVING, FAITHFUL, AND DEVOTED WIFE

THIS BOOK IS DEDICATED.

PREFACE.

The progress of The Salvation Army in its work amongst the poor and lost of many lands has compelled me to face the problems which are more or less hopefully considered in the following pages. The grim necessities of a huge Campaign carried on for many years against the evils which lie at the root of all the miseries of modern life, attacked in a thousand and one forms by a thousand and one lieutenants, have led me step by step to contemplate as a possible solution of at least some of those problems the Scheme of Social Selection and Salvation which I have here set forth.

When but a mere child the degradation and helpless misery of the poor Stockingers of my native town, wandering gaunt and hunger-stricken through the streets droning out their melancholy ditties, crowding the Union or toiling like galley slaves on relief works for a bare subsistence, kindled in my heart yearnings to help the poor which have continued to this day and which have had a powerful influence on my whole life. At last I may be going to see my longings to help the workless realised. I think I am.

The commiseration then awakened by the misery of this class has been an impelling force which has never ceased to make itself felt during forty years of active service in the salvation of men. During this time I am thankful that I have been able, by the good hand of God upon me, to do something in mitigation of the miseries of this class, and to bring not only heavenly hopes and earthly gladness to the hearts of multitudes of these wretched crowds, but also many material blessings, including such commonplace things as food, raiment, home, and work, the parent of so many other temporal benefits. And thus many poor creatures have proved Godliness to be "profitable unto all things, having the promise of the life that now is as well as of that which is to come."

These results have been mainly attained by spiritual means. I have boldly asserted that whatever his peculiar character or circumstances might be, if the prodigal would come home to his Heavenly Father, he would find enough and to spare in the Father's house to supply all his need both for this world and the next; and I have known thousands, nay, I can say tens of thousands, who have literally proved this to be true, having, with little or no temporal assistance, come out of the darkest depths of destitution, vice and crime, to be happy and honest citizens and true sons and servants of God.

And yet all the way through my career I have keenly felt the remedial measures usually enunciated in Christian programmes and ordinarily employed by Christian philanthropy to be lamentably inadequate for any effectual dealing with the despairing miseries of these outcast classes. The rescued are appallingly few—a ghastly minority compared with the multitudes who struggle and sink in the open-mouthed abyss. Alike, therefore, my humanity and my Christianity, if I may speak of them in any way as separate one from the other, have cried out for some more comprehensive method of reaching and saving the perishing crowds.

No doubt it is good for men to climb unaided out of the whirlpool on to the rock of deliverance in the very presence of the temptations which have hitherto mastered them, and to maintain a footing there with the same billows of temptation washing over them. But, alas! with many this seems to be literally impossible. That decisiveness of character, that moral nerve which takes hold of the rope thrown for the rescue and keeps its hold amidst all the resistances that have to be encountered, is wanting. It is gone. The general wreck has shattered and disorganised the whole man.

Alas, what multitudes there are around us everywhere, many known to my readers personally, and any number who may be known to them by a very short walk from their own dwellings, who are in this very plight! Their vicious habits and destitute circumstances make it certain that, without some kind of extraordinary help, they must hunger and sin, and sin and hunger, until, having multiplied their kind, and filled up the measure of their miseries, the gaunt fingers of death will close upon them and terminate their wretchedness. And all this will happen this very winter in the midst of the unparalleled wealth, and civilisation, and philanthropy of this professedly most Christian land.

Now, I propose to go straight for these sinking classes, and in doing so shall continue to aim at the heart. I still prophesy the uttermost disappointment unless that citadel is reached. In proposing to add one more to the methods I have already put into operation to this end, do not let it be supposed that I am the less dependent upon the old plans, or that I seek anything short of the old conquest. If we help the man it is in order that we may change him. The builder who should elaborate his design and erect his house and risk his reputation without burning his bricks would be pronounced a failure and a fool. Perfection of architectural beauty, unlimited expenditure of capital, unfailing watchfulness of his labourers, would avail him nothing if the bricks were merely unkilned clay. Let him kindle a fire. And so here I see the folly of hoping to accomplish anything abiding, either in the circumstances or the morals of these hopeless classes, except there be a change effected in the whole man as well as in his surroundings. To this everything I hope to attempt will tend. In many cases I shall succeed, in some I shall fail; but even in failing of this my ultimate design, I shall at least benefit the bodies, if not the souls, of men; and if I do not save the fathers, I shall make a better chance for the children.

It will be seen, therefore, that in this or in any other development that may follow, I have no intention to depart in the smallest degree from the

A

main principles on which I have acted in the past.   My only hope for the permanent deliverance of mankind from misery, either in this world or the next, is the regeneration or remaking of the individual by the power of the Holy Ghost through Jesus Christ.   But in providing for the relief of temporal misery I reckon that I am only making it easy where it is now difficult, and possible where it is now all but impossible, for men and women to find their way to the Cross of our Lord Jesus Christ.

That I have confidence in my proposals goes without saying.   I believe they will work.   In miniature many of them are working already.   But I do not claim that my Scheme is either perfect in its details or complete in the sense of being adequate to combat all forms of the gigantic evils against which it is in the main directed.   Like other human things it must be perfected through suffering.   But it is a sincere endeavour to do something, and to do it on principles which can be instantly applied and universally developed.   Time, experience, criticism, and, above all, the guidance of God will enable us, I hope, to advance on the lines here laid down to a true and practical application of the words of the Hebrew Prophet : "Loose the bands of wickedness ; undo the heavy burdens ; let the oppressed go free; break every yoke ; deal thy bread to the hungry; bring the poor that are cast out to thy house.   When thou seest the naked cover him and hide not thyself from thine own flesh.   Draw out thy soul to the hungry—Then they that be of thee shall build the old waste places *and* Thou shalt raise up the foundations of many generations.

To one who has been for nearly forty years indissolubly associated with me in every undertaking I owe much of the inspiration which has found expression in this book.   It is probably difficult for me to fully estimate the extent to which the splendid benevolence and unbounded sympathy of her character have pressed me forward in the life-long service of man, to which we have devoted both ourselves and our children.   It will be an ever green and precious memory to me that amid the ceaseless suffering of a dreadful malady my dying wife found relief in considering and developing the suggestions for the moral and social and spiritual blessing of the people which are here set forth, and I do thank God she was taken from me only when the book was practically complete and the last chapters had been sent to the press.

In conclusion, I have to acknowledge the services rendered to me in preparing this book by Officers under my command. There could be no hope of carrying out any part of it, but for the fact that so many thousands are ready at my call and under my direction to labour to the very utmost of their strength for the salvation of others without the hope of earthly reward. Of the practical common sense, the resource, the readiness for every form of usefulness of those Officers and Soldiers, the world has no conception. Still less is it capable of understanding the height and depth of their self-sacrificing devotion to God and the poor.

I have also to acknowledge valuable literary help from a friend of the poor, who, though not in any way connected with the Salvation Army, has the deepest sympathy with its aims and is to a large extent in harmony with its principles. Without such assistance I should probably have found it—overwhelmed as I already am with the affairs of a world-wide enterprise—extremely difficult, if not impossible, to have presented these proposals for which I am alone responsible in so complete a form, at any rate at this time. I have no doubt that if any substantial part of my plan is successfully carried out he will consider himself more than repaid for the services so ably rendered.

WILLIAM BOOTH.

International Headquarters of

The Salvation Army,

London, E.C., *October*, 1890.

CONTENTS.

PART I. — THE DARKNESS.

PART II.—DELIVERANCE.

CHAPTER I.

A STUPENDOUS UNDERTAKING.

CHAPTER II.

TO THE RESCUE!—THE CITY COLONY.

CHAPTER III.

TO THE COUNTRY!—THE FARM COLONY.

CHAPTER IV.

NEW BRITAIN.—THE COLONY OVER SEA.

CHAPTER V.

MORE CRUSADES.

CHAPTER VI.

ASSISTANCE IN GENERAL.

CHAPTER VII.

CAN IT BE DONE, AND HOW?

CHAPTER VIII.

IN DARKEST ENGLAND

PART I.—THE DARKNESS.

CHAPTER I.
WHY "DARKEST ENGLAND"?

This summer the attention of the civilised world has been arrested by the story which Mr. Stanley has told of "Darkest Africa" and his journeyings across the heart of the Lost Continent. In all that spirited narrative of heroic endeavour, nothing has so much impressed the imagination, as his description of the immense forest, which offered an almost impenetrable barrier to his advance. The intrepid explorer, in his own phrase, "marched, tore, ploughed, and cut his way for one hundred and sixty days through this inner womb of the true tropical forest." The mind of man with difficulty endeavours to realise this immensity of wooded wilderness, covering a territory half as large again as the whole of France, where the rays of the sun never penetrate, where in the dark, dank air, filled with the steam of the heated morass, human beings dwarfed into pygmies and brutalised into cannibals lurk and live and die. Mr. Stanley vainly endeavours to bring home to us the full horror of that awful gloom. He says:

Take a thick Scottish copse dripping with rain; imagine this to be a mere undergrowth nourished under the impenetrable shade of ancient trees ranging from 100 to 180 feet high; briars and thorns abundant; lazy creeks meandering through the depths of the jungle, and sometimes a deep affluent of a great river. Imagine this forest and jungle in all stages of decay and growth, rain pattering on you every other day of the year; an impure atmosphere with its dread consequences, fever and dysentery; gloom throughout the day and darkness almost palpable throughout the night; and then if you can imagine such a forest extending the entire distance from Plymouth to Peterhead, you will have a fair idea of some of the inconveniences endured by us in the Congo forest.

The denizens of this region are filled with a conviction that the forest is endless—interminable. In vain did Mr. Stanley and his companions endeavour to convince them that outside the dreary wood were to be found sunlight, pasturage and peaceful meadows.

They replied in a manner that seemed to imply that we must be strange creatures to suppose that it would be possible for any world to exist save their illimitable forest. "No," they replied, shaking their heads compassionately, and pitying our absurd questions, "all like this," and they moved their hands sweepingly to illustrate that the world was all alike, nothing but trees, trees and trees—great trees rising as high as an arrow shot to the sky, lifting their crowns intertwining their branches, pressing and crowding one against the other, until neither the sunbeam nor shaft of light can penetrate it.

"We entered the forest," says Mr. Stanley, "with confidence ; forty pioneers in front with axes and bill hooks to clear a path through the obstructions, praying that God and good fortune would lead us." But before the conviction of the forest dwellers that the forest was without end, hope faded out of the hearts of the natives of Stanley's company. The men became sodden with despair, preaching was useless to move their brooding sullenness, their morbid gloom.

The little religion they knew was nothing more than legendary lore, and in their memories there dimly floated a story of a land which grew darker and darker as one travelled towards the end of the earth and drew nearer to the place where a great serpent lay supine and coiled round the whole world. Ah! then the ancients must have referred to this, where the light is so ghastly, and the woods are endless, and are so still and solemn and grey ; to this oppressive loneliness, amid so much life, which is so chilling to the poor distressed heart ; and the horror grew darker with their fancies ; the cold of early morning, the comfortless grey of dawn, the dead white mist, the ever-dripping tears of the dew, the deluging rains, the appalling thunder bursts and the echoes, and the wonderful play of the dazzling lightning. And when the night comes with its thick palpable darkness, and they lie huddled in their damp little huts, and they hear the tempest overhead, and the howling of the wild winds, the grinding and groaning of the storm-tost trees, and the dread sounds of the falling giants, and the shock of the trembling earth which sends their hearts with fitful leaps to their throats, and the roaring and a rushing as of a mad overwhelming sea— oh, then the horror is intensified ! When the march has begun once again, and the files are slowly moving through the woods, they renew their morbid broodings, and ask themselves : How long is this to last ? Is the joy of life to end thus ? Must we jog on day after day in this cheerless gloom and this joyless duskiness, until we stagger and fall and rot among the toads ? Then they disappear into the woods by twos, and threes, and sixes ; and after the caravan has passed they return by the trail, some to reach Yambuya and upset the young officers with their tales of woe and war ; some to fall sobbing under a spear-thrust ; some to wander and stray in the dark mazes of the woods, hopelessly lost ; and some to be carved for the cannibal feast. And those who remain compelled to it by fears of greater danger, mechanically march on, a prey to dread and weakness.

That is the forest. But what of its denizens ? They are comparatively few ; only some hundreds of thousands living in small tribes from ten to thirty miles apart, scattered over an area on which ten thousand million trees put out the sun from a region four times as wide as Great Britain. Of these pygmies there are two kinds; one a very degraded specimen with ferretlike eyes, close-set nose, more nearly approaching the baboon than was supposed to be possible, but very human; the other very handsome, with frank open innocent features, very prepossessing. They are quick and intelligent, capable of deep affection and gratitude, showing remarkable industry and patience. A pygmy boy of eighteen worked with consuming zeal; time with him was too precious to waste in talk. His mind seemed ever concentrated on work. Mr. Stanley said :

" When I once stopped him to ask him his name, his face seemed to say, ' Please don't stop me. I must finish my task.'

" All alike, the baboon variety and the handsome innocents, are cannibals. They are possessed with a perfect mania for meat. We were obliged to bury our dead in the river, lest the bodies should be exhumed and eaten, even when they had died from smallpox."

Upon the pygmies and all the dwellers of the forest has descended a devastating visitation in the shape of the ivory raiders of civilisation. The race that wrote the Arabian Nights, built Bagdad and Granada, and invented Algebra, sends forth men with the hunger for gold in their hearts, and Enfield muskets in their hands, to plunder and to slay. They exploit the domestic affections of the forest dwellers in order to strip them of all they possess in the world. That has been going on for years. It is going on to-day. It has come to be regarded as the natural and normal law of existence. Of the religion of these hunted pygmies Mr. Stanley tells us nothing, perhaps because there is nothing to tell. But an earlier traveller, Dr. Kraff, says that one of these tribes, by name Doko, had some notion of a Supreme Being, to whom, under the name of Yer, they sometimes addressed prayers in moments of sadness or terror. In these prayers they say; "Oh Yer, if Thou dost really exist why dost Thou let us be slaves ? We ask not for food or clothing, for we live on snakes, ants, and mice. Thou hast made us, wherefore dost Thou let us be trodden down ?"

It is a terrible picture, and one that has engraved itself deep on the heart of civilisation. But while brooding over the awful presentation of life as it exists in the vast African forest, it seemed to me only too vivid a picture of many parts of our own land. As there is a darkest Africa is there not also a darkest England ? Civilisation, which can breed its own barbarians, does it not also breed its own pygmies ? May we not find a parallel at our own

doors, and discover within a stone's throw of our cathedrals and
palaces similar horrors to those which Stanley has found existing
in the great Equatorial forest?

The more the mind dwells upon the subject, the closer the analogy
appears. The ivory raiders who brutally traffic in the unfortunate
denizens of the forest glades, what are they but the publicans who
flourish on the weakness of our poor? The two tribes of savages,
the human baboon and the handsome dwarf, who will not speak
lest it impede him in his task, may be accepted as the two
varieties who are continually present with us—the vicious, lazy
lout, and the toiling slave. They, too, have lost all faith of life
being other than it is and has been. As in Africa, it is all trees,
trees, trees with no other world conceivable; so is it here—it is all
vice and poverty and crime. To many the world is all slum, with
the Workhouse as an intermediate purgatory before the grave. And
just as Mr. Stanley's Zanzibaris lost faith, and could only be induced
to plod on in brooding sullenness of dull despair, so the most of our
social reformers, no matter how cheerily they may have started off,
with forty pioneers swinging blithely their axes as they force their
way into the wood, soon become depressed and despairing. Who
can battle against the ten thousand million trees? Who can hope to
make headway against the innumerable adverse conditions which
doom the dweller in Darkest England to eternal and immutable
misery? What wonder is it that many of the warmest hearts and
enthusiastic workers feel disposed to repeat the lament of the old
English chronicler, who, speaking of the evil days which fell upon
our forefathers in the reign of Stephen, said " It seemed to them as
if God and his Saints were dead."

An analogy is as good as a suggestion; it becomes wearisome
when it is pressed too far. But before leaving it, think for a moment
how close the parallel is, and how strange it is that so much interest
should be excited by a narrative of human squalor and human
heroism in a distant continent, while greater squalor and heroism
not less magnificent may be observed at our very doors.

The Equatorial Forest traversed by Stanley resembles that Darkest
England of which I have to speak, alike in its vast extent—both stretch,
in Stanley's phrase, " as far as from Plymouth to Peterhead ;" its monotonous darkness, its malaria and its gloom, its dwarfish de-humanized
inhabitants, the slavery to which they are subjected, their privations
and their misery. That which sickens the stoutest heart, and causes

many of our bravest and best to fold their hands in despair, is the
apparent impossibility of doing more than merely to peck at the
outside of the endless tangle of monotonous undergrowth ; to let
light into it, to make a road clear through it, that shall not be immediately choked up by the ooze of the morass and the luxuriant para-
sitical growth of the forest—who dare hope for that ?   At present,
alas, it would seem as though no one dares even to hope !   It is the
great Slough of Despond of our time.

And what a slough it is no man can gauge who has not waded
therein, as some of us have done, up to the very neck for long years.
Talk about Danté's Hell, and all the horrors and cruelties of the
torture-chamber of the lost !   The man who walks with open eyes
and with bleeding heart through the shambles of our civilisation
needs no such fantastic images of the poet to teach him horror.
Often and often, when I have seen the young and the poor and the
helpless go down before my eyes into the morass, trampled underfoot
by beasts of prey in human shape that haunt these regions, it seemed
as if God were no longer in His world, but that in His stead reigned
a fiend, merciless as Hell, ruthless as the grave.   Hard it is, no doubt,
to read in Stanley's pages of the slave-traders coldly arranging for
the surprise of a village, the capture of the inhabitants, the massacre
of those who resist, and the violation of all the women; but the stony
streets of London, if they could but speak, would tell of tragedies as
awful, of ruin as complete, of ravishments as horrible, as if we were
in Central Africa; only the ghastly devastation is covered, corpse-
like, with the artificialities and hypocrisies of modern civilisation.

The lot of a negress in the Equatorial Forest is not, perhaps, a very
happy one, but is it so very much worse than that of many a pretty
orphan girl in our Christian capital ?   We talk about the brutalities
of the dark ages, and we profess to shudder as we read in books of
the shameful exaction of the rights of feudal superior.   And yet here,
beneath our very eyes, in our theatres, in our restaurants, and in many
other places, unspeakable though it be but to name it, the same hideous
abuse flourishes unchecked.   A young penniless girl, if she be pretty,
is often hunted from pillar to post by her employers, confronted always
by the alternative—Starve or Sin.   And when once the poor girl has
consented to buy the right to earn her living by the sacrifice of her
virtue, then she is treated as a slave and an outcast by the
very men who have ruined her.   Her word becomes unbeliev-
able, her life an ignominy, and she is swept downward

ever downward, into the bottomless perdition of prostitution. But there, even in the lowest depths, excommunicated by Humanity and outcast from God, she is far nearer the pitying heart of the One true Saviour than all the men who forced her down, aye, and than all the Pharisees and Scribes who stand silently by while these fiendish wrongs are perpetrated before their very eyes.

The blood boils with impotent rage at the sight of these enormities, callously inflicted, and silently borne by these miserable victims. Nor is it only women who are the victims, although their fate is the most tragic. Those firms which reduce sweating to a fine art, who systematically and deliberately defraud the workman of his pay, who grind the faces of the poor, and who rob the widow and the orphan, and who for a pretence make great professions of public-spirit and philanthropy, these men nowadays are sent to Parliament to make laws for the people. The old prophets sent them to Hell—but we have changed all that. They send their victims to Hell, and are rewarded by all that wealth can do to make their lives comfortable. Read the House of Lords' Report on the Sweating System, and ask if any African slave system, making due allowance for the superior civili-sation, and therefore sensitiveness, of the victims, reveals more misery.

Darkest England, like Darkest Africa, reeks with malaria. The foul and fetid breath of our slums is almost as poisonous as that of the African swamp. Fever is almost as chronic there as on the Equator. Every year thousands of children are killed off by what is called defects of our sanitary system. They are in reality starved and poisoned, and all that can be said is that, in many cases, it is better for them that they were taken away from the trouble to come.

Just as in Darkest Africa it is only a part of the evil and misery that comes from the superior race who invade the forest to enslave and massacre its miserable inhabitants, so with us, much of the misery of those whose lot we are considering arises from their own habits. Drunkenness and all manner of uncleanness, moral and physical, abound. Have you ever watched by the bedside of a man in delirium tremens ? Multiply the sufferings of that one drunkard by the hundred thousand, and you have some idea of what scenes are being witnessed in all our great cities at this moment. As in Africa streams intersect the forest in every direction, so the gin-shop stands at every corner with its River of the Water of Death flowing seventeen hours out of the twenty-four for the destruction of the people. A population sodden with drink, steeped in vice,

LIGHT BEYOND.   15

eaten up by every social and physical malady, these are the denizens
of Darkest England amidst whom my life has been spent, and to
whose rescue I would now summon all that is best in the manhood
and womanhood of our land.

But this book is no mere lamentation of despair. For Darkest
England, as for Darkest Africa, there is a light beyond. I think
I see my way out, a way by which these wretched ones may escape
from the gloom of their miserable existence into a higher and happier
life. Long wandering in the Forest of the Shadow of Death at our
doors, has familiarised me with its horrors ; but while the realisation
is a vigorous spur to action it has never been so oppressive as to
extinguish hope. Mr. Stanley never succumbed to the terrors which
oppressed his followers. He had lived in a larger life, and knew
that the forest, though long, was not interminable. Every step
forward brought him nearer his destined goal, nearer to the light of
the sun, the clear sky, and the rolling uplands of the grazing land
Therefore he did not despair. The Equatorial Forest was, after all,
a mere corner of one quarter of the world. In the knowledge of the
light outside, in the confidence begotten by past experience of successful endeavour, he pressed forward ; and when the 160 days'
struggle was over, he and his men came out into a pleasant place
where the land smiled with peace and plenty, and their hardships
and hunger were forgotten in the joy of a great deliverance.

So I venture to believe it will be with us. But the end is not yet.
We are still in the depths of the depressing gloom. It is in no spirit
of light-heartedness that this book is sent forth into the world.
The magnitude of the evils and the difficulty of dealing with them
are immense.

If this were the first time that this wail of hopeless misery had
sounded on our ears the matter would have been less serious. It is
because we have heard it so often that the case is so desperate.
The exceeding bitter cry of the disinherited has become to be as
familiar in the ears of men as the dull roar of the streets or as the
moaning of the wind through the trees. And so it rises unceasing,
year in and year out, and we are too busy or too idle, too indifferent
or too selfish, to spare it a thought. Only now and then, on rare occa-
sions, when some clear voice is heard giving more articulate utterance
to the miseries of the miserable men, do we pause in the regular routine
of our daily duties, and shudder as we realise for one brief moment
what life means to the inmates of the Slums. But one of the grimmest
social problems of our time should be sternly faced, not with a view

to the generation of profitless emotion, but with a view to its solution.

Is it not time? There is, it is true, an audacity in the mere suggestion that the problem is not insoluble that is enough to take away the breath. But can nothing be done? If, after full and exhaustive consideration, we come to the deliberate conclusion that nothing can be done, and that it is the inevitable and inexorable destiny of thousands of Englishmen to be brutalised into worse than beasts by the condition of their environment, so be it. But if, on the contrary, we are unable to believe that this " awful slough," which engulfs the manhood and womanhood of generation after generation, is incapable of removal ; and if the heart and intellect of mankind alike revolt against the fatalism of despair, then, indeed, it is time, and high time, that the question were faced in no mere dilettante spirit, but with a resolute determination to make an end of the crying scandal of our age.

What a satire it is upon our Christianity and our civilisation, that the existence of these colonies of heathens and savages in the heart of our capital should attract so little attention ! It is no better than a ghastly mockery—theologians might use a stronger word—to call by the name of One who came to seek and to save that which was lost those Churches which in the midst of lost multitudes either sleep in apathy or display a fitful interest in a chasuble. Why all this apparatus of temples and meeting-houses to save men from perdition in a world which is to come, while never a helping hand is stretched out to save them from the inferno of their present life? Is it not time that, forgetting for a moment their wranglings about the infinitely little or infinitely obscure, they should concentrate all their energies on a united effort to break this terrible perpetuity of perdition, and to rescue some at least of those for whom they profess to believe their Founder came to die?

Before venturing to define the remedy, I begin by describing the malady. But even when presenting the dreary picture of our social ills, and describing the difficulties which confront us, I speak not in despondency but in hope. "I know in whom I have believed." I know, therefore do I speak. Darker England is but a fractional part of " Greater England." There is wealth enough abundantly to minister to its social regeneration so far as wealth can, if there be but heart enough to set about the work in earnest. And I hope and believe that the heart will not be lacking when once the problem is manfully faced, and the method of its solution plainly pointed out.

CHAPTER II.

THE SUBMERGED TENTH.

In setting forth the difficulties which have to be grappled with, I shall endeavour in all things to understate rather than overstate my case. I do this for two reasons: first, any exaggeration would create a reaction; and secondly, as my object is to demonstrate the practicability of solving the problem, I do not wish to magnify its dimensions. In this and in subsequent chapters I hope to convince those who read them that there is no overstraining in the representation of the facts, and nothing Utopian in the presentation of remedies. I appeal neither to hysterical emotionalists nor headlong enthusiasts; but having tried to approach the examination of this question in a spirit of scientific investigation, I put forth my proposals with the view of securing the support and co-operation of the sober, serious, practical men and women who constitute the saving strength and moral backbone of the country. I fully admit that there is much that is lacking in the diagnosis of the disease, and, no doubt, in this first draft of the prescription there is much room for improvement, which will come when we have the light of fuller experience. But with all its drawbacks and defects, I do not hesitate to submit my proposals to the impartial judgment of all who are interested in the solution of the social question as an immediate and practical mode of dealing with this, the greatest problem of our time.

The first duty of an investigator in approaching the study of any question is to eliminate all that is foreign to the inquiry, and to concentrate his attention upon the subject to be dealt with. Here I may remark that I make no attempt in this book to deal with Society as a whole. I leave to others the formulation of ambitious programmes for the reconstruction of our entire social system; not because I may not desire its reconstruction, but because the elaboration of any plans which are more or less visionary and incapable of realisation for many years would stand in the way of the consideration of this Scheme for dealing with the most urgently pressing aspect of the question, which I hope may be put into operation at once.

In taking this course I am aware that I cut myself off from a wide and attractive field ; but as a practical man, dealing with sternly prosaic facts, I must confine my attention to that particular section of the problem which clamours most pressingly for a solution. Only one thing I may say in passing. There is nothing in my scheme which will bring it into collision either with Socialists of the State, or Socialists of the Municipality, with Individualists or Nationalists, or any of the various schools of thought in the great field of social economics—excepting only those anti-Christian economists who hold that it is an offence against the doctrine of the survival of the fittest to try to save the weakest from going to the wall, and who believe that when once a man is down the supreme duty of a self-regarding Society is to jump upon him. Such economists will naturally be disappointed with this book. I venture to believe that all others will find nothing in it to offend their favourite theories, but perhaps something of helpful suggestion which they may utilise hereafter.

What, then, is Darkest England ? For whom do we claim that "urgency" which gives their case priority over that of all other sections of their countrymen and countrywomen ?

I claim it for the Lost, for the Outcast, for the Disinherited of the World.

These, it may be said, are but phrases. Who are the Lost ? I reply, not in a religious, but in a social sense, the lost are those who have gone under, who have lost their foothold in Society, those to whom the prayer to our Heavenly Father, " Give us day by day our daily bread," is either unfulfilled, or only fulfilled by the Devil's agency : by the earnings of vice, the proceeds of crime, or the contribution enforced by the threat of the law.

But I will be more precise. The denizens in Darkest England, for whom I appeal, are (1) those who, having no capital or income of their own, would in a month be dead from sheer starvation were they exclusively dependent upon the money earned by their own work ; and (2) those who by their utmost exertions are unable to attain the regulation allowance of food which the law prescribes as indispensable even for the worst criminals in our gaols.

I sorrowfully admit that it would be Utopian in our present social arrangements to dream of attaining for every honest Englishman a gaol standard of all the necessaries of life. Some time, perhaps, we may venture to hope that every honest worker on English soil will always be as warmly clad, as healthily housed, and as regularly fed as our criminal convicts—but that is not yet.

Neither is it possible to hope for many years to come that human beings generally will be as well cared for as horses. Mr. Carlyle long ago remarked that the four-footed worker has already got all that this two-handed one is clamouring for : "There are not many horses in England, able and willing to work, which have not due food and lodging and go about sleek coated, satisfied in heart." You say it is impossible ; but, said Carlyle, "The human brain, looking at these sleek English horses, refuses to believe in such impossibility for English men." Nevertheless, forty years have passed since Carlyle said that, and we seem to be no nearer the attainment of the four-footed standard for the two-handed worker. "Perhaps it might be nearer realisation," growls the cynic, "if we could only produce men according to demand, as we do horses, and promptly send them to the slaughter-house when past their prime"—which, of course, is not to be thought of.

What, then, is the standard towards which we may venture to aim with some prospect of realisation in our time? It is a very humble one, but if realised it would solve the worst problems of modern Society.

It is the standard of the London Cab Horse.

When in the streets of London a Cab Horse, weary or careless or stupid, trips and falls and lies stretched out in the midst of the traffic, there is no question of debating how he came to stumble before we try to get him on his legs again. The Cab Horse is a very real illustration of poor broken-down humanity; he usually falls down because of overwork and underfeeding. If you put him on his feet without altering his conditions, it would only be to give him another dose of agony ; but first of all you'll have to pick him up again. It may have been through overwork or underfeeding, or it may have been all his own fault that he has broken his knees and smashed the shafts, but that does not matter. If not for his own sake, then merely in order to prevent an obstruction of the traffic, all attention is concentrated upon the question of how we are to get him on his legs again. The load is taken off, the harness is unbuckled, or, if need be, cut, and everything is done to help him up. Then he is put in the shafts again and once more restored to his regular round of work. That is the first point. The second is that every Cab Horse in London has three things ; a shelter for the night, food for its stomach, and work allotted to it by which it can earn its corn.

These are the two points of the Cab Horse's Charter. When he is down he is helped up, and while he lives he has food, shelter and work. That, although a humble standard, is at present absolutely unattainable by millions—literally by millions—of our fellow-men and women in this country. Can the Cab Horse Charter be gained for human beings? I answer, yes. The Cab Horse standard can be attained on the Cab Horse terms. If you get your fallen fellow on his feet again, Docility and Discipline will enable you to reach the Cab Horse ideal, otherwise it will remain unattainable. But Docility seldom fails where Discipline is intelligently maintained. Intelligence is more frequently lacking to direct, than obedience to follow direction. At any rate it is not for those who possess the intelligence to despair of obedience, until they have done their part. Some, no doubt, like the bucking horse that will never be broken in, will always refuse to submit to any guidance but their own lawless will. They will remain either the Ishmaels or the Sloths of Society. But man is naturally neither an Ishmael nor a Sloth.

The first question, then, which confronts us is, what are the dimensions of the Evil? How many of our fellow-men dwell in this Darkest England? How can we take the census of those who have fallen below the Cab Horse standard to which it is our aim to elevate the most wretched of our countrymen?

The moment you attempt to answer this question, you are confronted by the fact that the Social Problem has scarcely been studied at all scientifically. Go to Mudie's and ask for all the books that have been written on the subject, and you will be surprised to find how few there are. There are probably more scientific books treating of diabetes or of gout than there are dealing with the great social malady which eats out the vitals of such numbers of our people. Of late there has been a change for the better. The Report of the Royal Commission on the Housing of the Poor, and the Report of the Committee of the House of Lords on Sweating, represent an attempt at least to ascertain the facts which bear upon the Condition of the People question. But, after all, more minute, patient, intelligent observation has been devoted to the study of Earthworms, than to the evolution, or rather the degradation, of the Sunken Section of our people.   Here and there in the immense field individual workers make notes, and occasionally emit a wail of despair, but where is there any attempt even so much as to take the first preliminary step of counting those who have gone under ?

One book there is, and so far as I know at present, only one, which even attempts to enumerate the destitute.   In his " Life and Labour in the East of London," Mr. Charles Booth attempts to form some kind of an idea as to the numbers of those with whom we have to deal.   With a large staff of assistants, and provided with all the facts in possession of the School Board Visitors, Mr. Booth took an industrial census of East London.   This district, which comprises Tower Hamlets, Shoreditch, Bethnal Green and Hackney, contains a population of 908,000; that is to say, less than one-fourth of the population of London.

How do his statistics work out ?   If we estimate the number of the poorest class in the rest of London as being twice as numerous as those in the Eastern District, instead of being thrice as numerous, as they would be if they were calculated according to the population in the same proportion, the following is the result :—

	East London	Estimate for rest of London.	Total
PAUPERS			
Inmates of Workhouses, Asylums, and Hospitals	17,000 ...	34,000 ...	51,000
HOMELESS			
Loafers, Casuals, and some Criminals	11,000 ...	22,000 ...	33,000
STARVING			
Casual earnings between 18s. per week and chronic want	100,000 ...	200,000 ...	300,000
THE VERY POOR.			
Intermittent earnings 18s. to 21s. per week	74,000 ...	148,000 ...	222,000
Small regular earnings 18s. to 21s. per week	129,000 ...	258,000 ...	387,000
	331,000	662,000	993,000
Regular wages, artizans, etc., 22s. to 30s. per week	377,000		
Higher class labour, 30s. to 50s. per week	121,000		
Lower middle class, shopkeepers, clerks, etc.	34,000		
Upper middle class (servant keepers)	45,000		
	908,000		

It may be admitted that East London affords an exceptionally bad district from which to generalise for the rest of the country. Wages are higher in London than elsewhere, but so is rent, and the number of the homeless and starving is greater in the human warren at the East End. There are 31 millions of people in Great Britain, exclusive of Ireland. If destitution existed everywhere in East London proportions, there would be 31 times as many homeless and starving people as there are in the district round Bethnal Green.

But let us suppose that the East London rate is double the average for the rest of the country. That would bring out the following figures :—

HOUSELESS

	East London.		United Kingdom.
Loafers, Casuals, and some Criminals ...	11,000	......	165,500
STARVING			
Casual earnings or chronic want...	...100,000	......	1,550,000
Total Houseless and Starving	...111,000	......	1,715,500
In Workhouses, Asylums, &c.	... 17,000	......	190,000
	128,000		1,905,500

Of those returned as homeless and starving, 870,000 were in receipt of outdoor relief.

To these must be added the inmates of our prisons. In 1889, 174,779 persons were received in the prisons, but the average number in prison at any one time did not exceed 60,000. The figures, as given in the Prison Returns, are as follows :—

In Convict Prisons	...	...	...	...	... 11,660
In Local Prisons ...	...	...	...	...	... 20,883
In Reformatories ...	...	...	...	...	... 1,270
In Industrial Schools	...	...	...	...	... 21,413
Criminal Lunatics	...	...	...	...	... 910
					56,136

Add to this the number of indoor paupers and lunatics (excluding criminals) 78,966—and we have an army of nearly two millions belonging to the submerged classes. To this there must be added, at the very least, another million, representing those dependent upon the criminal, lunatic and other classes, not enumerated here, and the more or less helpless of the class immediately above the houseless and starving. This brings my total to three millions, or, to put it roughly to one-tenth of the population. According to Lord Brabazon and Mr. Samuel Smith, "between two and three millions of our population are always pauperised and degraded." Mr. Chamberlain says there is a " population equal to that of the metropolis,"—that is, between four and five millions—" which has remained constantly in a state of abject destitution and misery." Mr. Giffen is more moderate. The submerged class, according to him, comprises one in five of manual labourers, six in 100 of the population. Mr. Giffen does not add the third million which is living on the border line. Between Mr. Chamberlain's four millions and a half, and Mr. Giffen's 1,800,000, I am content to take three millions as representing the total strength of the destitute army.

Darkest England, then, may be said to have a population about equal to that of Scotland. Three million men, women, and children, a vast despairing multitude in a condition nominally free, but really enslaved ;—these it is whom we have to save.

It is a large order. England emancipated her negroes sixty years ago, at a cost of £40,000,000, and has never ceased boasting about it since. But at our own doors, from " Plymouth to Peterhead," stretches this waste Continent of humanity—three million human beings who are enslaved—some of them to taskmasters as merciless as any West Indian overseer, all of them to destitution and despair.

Is anything to be done with them ? Can anything be done for them ? Or is this million-headed mass to be regarded as offering a problem as insoluble as that of the London sewage, which, feculent and festering, swings heavily up and down the basin of the Thames with the ebb and flow of the tide ?

This Submerged Tenth—is it, then, beyond the reach of the nine-tenths in the midst of whom they live, and around whose homes they rot and die ? No doubt, in every large mass of human beings there will be some incurably diseased in morals and in body, some for whom nothing can be done, some of whom even the optimist must despair, and for whom he can prescribe nothing‧ but the beneficently stern restraints of an asylum or a gaol.

But is not one in ten a proportion scandalously high ? The Israelites of old set apart one tribe in twelve to minister to the Lord in the service of the Temple ; but must we doom one in ten of " God's Englishmen " to the service of the great Twin Devils—Destitution and Despair ?

CHAPTER III.

THE HOMELESS.

Darkest England may be described as consisting broadly of three circles, one within the other. The outer and widest circle is inhabited by the starving and the homeless, but honest, Poor. The second by those who live by Vice ; and the third and innermost region at the centre is peopled by those who exist by Crime. The whole of the three circles is sodden with Drink. Darkest England has many more public-houses than the Forest of the Aruwimi has rivers, of which Mr. Stanley sometimes had to cross three in half-an-hour.

The borders of this great lost land are not sharply defined. They are continually expanding or contracting. Whenever there is a period of depression in trade, they stretch ; when prosperity returns, they contract. So far as individuals are concerned, there are none among the hundreds of thousands who live upon the out-skirts of the dark forest who can truly say that they or their children are secure from being hopelessly entangled in its labyrinth. The death of the bread-winner, a long illness, a failure in the City, or any one of a thousand other causes which might be named, will bring within the first circle those who at present imagine themselves free from all danger of actual want. The death-rate in Darkest England is high. Death is the great gaol-deliverer of the captives. But the dead are hardly in the grave before their places are taken by others. Some escape, but the majority, their health sapped by their surroundings, become weaker and weaker, until at last they fall by the way, perishing without hope at the very doors of the palatial mansions which, may-be, some of them helped to build.

Some seven years ago a great outcry was made concerning the Housing of the Poor. Much was said, and rightly said—it could not be said too strongly—concerning the disease-breeding, manhood-destroying character of many of the tenements in which the poor herd in our large cities. But there is a depth below that of the dweller in the slums. It is that of the dweller in the street, who has not even a lair in the slums which he can call his own. The houseless Out-of-Work is in one respect at least like Him of whom it was said, " Foxes have holes, and birds of the air have nests, but the Son of Man hath not where to lay His head."

The existence of these unfortunates was somewhat rudely forced upon the attention of Society in 1887, when Trafalgar Square became the camping ground of the Homeless Outcasts of London. Our Shelters have done something, but not enough, to provide for the outcasts, who this night and every night are walking about the streets, not knowing where they can find a spot on which to rest their weary frames.

Here is the return of one of my Officers who was told off this summer to report upon the actual condition of the Homeless who have no roof to shelter them in all London :—

There are still a large number of Londoners and a considerable percentage of wanderers from the country in search of work, who find themselves at night-fall destitute. These now betake themselves to the seats under the plane trees on the Embankment. Formerly they endeavoured to occupy all the seats, but the lynx-eyed Metropolitan Police declined to allow any such proceedings, and the dossers, knowing the invariable kindness of the City Police, made tracks for that portion of the Embankment which, lying east of the Temple, comes under the control of the Civic Fathers. Here, between the Temple and Blackfriars, I found the poor wretches by the score; almost every seat contained its full complement of six—some men, some women—all reclining in various postures and nearly all fast asleep. Just as Big Ben strikes two, the moon, flashing across the Thames and lighting up the stone work of the Embankment, brings into relief a pitiable spectacle. Here on the stone abutments, which afford a slight protection from the biting wind, are scores of men lying side by side, huddled together for warmth, and, of course, without any other covering than their ordinary clothing, which is scanty enough at the best. Some have laid down a few pieces of waste paper, by way of taking the chill off the stones, but the majority are too tired, even for that, and the nightly toilet of most consists of first removing the hat, swathing the head in whatever old rag may be doing duty as a handkerchief, and then replacing the hat.

The intelligent-looking elderly man, who was just fixing himself up on a seat, informed me that he frequently made that his night's abode. "You see," quoth he, "there's nowhere else so comfortable. I was here last night, and

Monday and Tuesday as well, that's four nights this week. I had no money for lodgings, couldn't earn any, try as I might. I've had one bit of bread to-day, nothing else whatever, and I've earned nothing to-day or yesterday; I had threepence the day before. Gets my living by carrying parcels, or minding horses, or odd jobs of that sort. You see I haven't got my health, that's where it is. I used to work on the London General Omnibus Company and after that on the Road Car Company, but I had to go to the infirmary with bronchitis and couldn't get work after that. What's the good of a man what's got bronchitis and just left the infirmary? Who'll engage him, I'd like to know? Besides, it makes me short of breath at times, and I can't do much. I'm a widower; wife died long ago. I have one boy, abroad, a sailor, but he's only lately started and can't help me. Yes! its very fair out here of nights, seats rather hard, but a bit of waste paper makes it a lot softer. We have women sleep here often, and children, too. They're very well conducted, and there's seldom many rows here, you see, because everybody's tired out. We're too sleepy to make a row."

Another party, a tall, dull, helpless-looking individual, had walked up from the country; would prefer not to mention the place. He had hoped to have obtained a hospital letter at the Mansion House so as to obtain a truss for a bad rupture, but failing, had tried various other places, also in vain, winding up minus money or food on the Embankment.

In addition to these sleepers, a considerable number walk about the streets up till the early hours of the morning to hunt up some job which will bring a copper into the empty exchequer, and save them from actual starvation. I had some conversation with one such, a stalwart youth lately discharged from the militia, and unable to get work.

"You see," said he, pitifully, "I don't know my way about like most of the London fellows. I'm so green, and don't know how to pick up jobs like they do. I've been walking the streets almost day and night these two weeks and can't get work. I've got the strength, though I shan't have it long at this rate. I only want a job. This is the third night running that I've walked the streets all night; the only money I get is by minding blacking-boys' boxes while they go into Lockhart's for their dinner. I got a penny yesterday at it, and twopence for carrying a parcel, and to-day I've had a penny. Bought a ha'porth of bread and a ha'penny mug of tea."

Poor lad! probably he would soon get into thieves' company, and sink into the depths, for there is no other means of living for many like him; it is starve or steal, even for the young. There are gangs of lad thieves in the low Whitechapel lodging-houses, varying in age from thirteen to fifteen, who live by thieving eatables and other easily obtained goods from shop fronts.

In addition to the Embankment, *al fresco* lodgings are found in the seats outside Spitalfields Church, and many homeless wanderers have their own little nooks and corners of resort in many sheltered yards, vans, etc., all over London. Two poor women I observed making their home in a shop door-way in Liverpool Street. Thus they manage in the summer; what it's like in winter time is terrible to think of. In many cases it means the pauper's grave, as in the case of a young woman who was wont to sleep in a van in Bedfordbury. Some men who were aware of her practice surprised her by dashing a bucket of water on her. The blow to her weak system caused illness, and the inevitable sequel— a coroner's jury came to the conclusion that the water only hastened her death, which was due, in plain English, to starvation.

The following are some statements taken down by the same Officer from twelve men whom he found sleeping on the Embankment on the nights of June 13th and 14th, 1890 :—

No. 1. "I've slept here two nights; I'm a confectioner by trade; I come from Dartford. I got turned off because I'm getting elderly. They can get young men cheaper, and I have the rheumatism so bad. I've earned nothing these two days; I thought I could get a job at Woolwich, so I walked there, but could get nothing. I found a bit of bread in the road wrapped up in a bit of newspaper. That did me for yesterday. I had a bit of bread and butter to-day. I'm 54 years old. When it's wet we stand about all night under the arches."

No. 2. "Been sleeping out three weeks all but one night; do odd jobs, mind horses, and that sort of thing. Earned nothing to-day, or shouldn't be here. Have had a pen'orth of bread to-day. That's all. Yesterday had some pieces given to me at a cook-shop. Two days last week had nothing at all from morning till-night. By trade I'm a feather-bed dresser, but it's gone out of fashion, and besides that, I've a cataract in one eye, and have lost the sight of it completely. I'm a widower, have one child, a soldier, at Dover. My last regular work was eight months ago, but the firm broke. Been doing odd jobs since."

No. 3. "I'm a tailor; have slept here four nights running. Can't get work. Been out of a job three weeks. If I can muster cash I sleep at a lodging-house in Vere Street, Clare Market. It was very wet last night. I left these seats and went to Covent Garden Market and slept under cover. There were about thirty of us. The police moved us on, but we went back as soon as they had gone. I've had a pen'orth of bread and pen'orth of soup during the last two days—often goes without altogether. There are women sleep out here. They are decent people, mostly charwomen and such like who can't get work."

No. 4. Elderly man; trembles visibly with excitement at mention of work; produces a card carefully wrapped in old newspaper, to the effect that Mr. J. R. is a member of the Trade Protection League. He is a waterside labourer; last job at that was a fortnight since. Has earned nothing for five days. Had a bit of bread this morning, but not a scrap since. Had a cup of tea and two slices of bread yesterday, and the same the day before; the deputy at a lodging house gave it to him. He is fifty years old, and is still damp from sleeping out in the wet last night.

No. 5. Sawyer by trade, machinery cut him out. Had a job, haymaking near Uxbridge. Had been on same job lately for a month; got 2s. 6d. a day. (Probably spent it in drink, seems a very doubtful worker.) Has been odd jobbing a long time, earned 2d. to-day, bought a pen'orth of tea and ditto of sugar (produces same from pocket) but can't get any place to make the tea; was hoping to get to a lodging house where he could borrow a teapot, but had no money. Earned nothing yesterday, slept at a casual ward; very poor place, get insufficient food, considering the labour. Six ounces of bread and a pint of skilly for breakfast, one ounce of cheese and six or seven ounces of bread for dinner (bread cut by guess). Tea same as breakfast,—no supper. For this you have to break 10 cwt. of stones, or pick 4 lbs. of oakum.

Number 6. Had slept out four nights running. Was a distiller by trade; been out four months; unwilling to enter into details of leaving, but it was his own fault. (Very likely; a heavy, thick, stubborn, and senseless-looking fellow, six feet high, thick neck, strong limbs, evidently destitute of ability.) Does odd jobs; earned 3d. for minding a horse, bought a cup of coffee and pen'orth of bread and butter. Has no money now. Slept under Waterloo Bridge last night.

No. 7. Good-natured looking man; one who would suffer and say nothing; clothes shining with age, grease, and dirt; they hang on his joints as on pegs; awful rags! I saw him endeavouring to walk. He lifted his feet very slowly and put them down carefully in evident pain. His legs are bad; been in infirmary several times with them. His uncle and grandfather were clergymen; both dead now. He was once in a good position in a money office, and afterwards in the London and County Bank for nine years. Then he went with an auctioneer who broke, and he was left ill, old, and without any trade. "A clerk's place," says he, "is never worth having, because there are so many of them, and once out you can only get another place with difficulty. I have a brother-in-law on the Stock Exchange, but he won't own me. Look at my clothes? Is it likely?"

No. 8. Slept here four nights running. Is a builder's labourer by trade, that is, a handy-man. Had a settled job for a few weeks which expired three weeks since. Has earned nothing for nine days. Then helped wash down a shop front and got 2s. 6d. for it. Does anything he can get. Is 46 years old. Earns about 2d. or 3d. a day at horse minding. A cup of tea and a bit of bread yesterday, and same to-day, is all he has had.

No. 9. A plumber's labourer (all these men who are somebody's "labourers" are poor samples of humanity, evidently lacking in grit, and destitute of ability to do any work which would mean decent wages). Judging from appearances, they will do nothing well. They are a kind of automaton, with the machinery rusty; slow, dull, and incapable. The man of ordinary intelligence leaves them in the rear. They could doubtless earn more even at odd jobs, but lack the energy. Of course, this means little food, exposure to weather, and increased incapability day by day. ("From him that hath not," etc.) Out of work through slackness, does odd jobs; slept here three nights running. Is a dock labourer when he *can* get work. Has 6d. an hour; works so many hours, according as he is wanted. Gets 2s., 3s., or 4s. 6d. a day. Has to work very hard for it. Casual ward life is also very hard, he says, for those who are not used to it, and there is not enough to eat. Has had to-day a pen'orth of bread, for minding a cab. Yesterday he spent 3½d. on a breakfast, and that lasted him all day. Age 25.

No. 10. Been out of work a month. Carman by trade. Arm withered, and cannot do work properly. Has slept here all the week; got an awful cold through the wet. Lives at odd jobs (they all do). Got sixpence yesterday for minding a cab and carrying a couple of parcels. Earned nothing to-day, but had one good meal; a lady gave it him. Has been walking about all day looking for work, and is tired out.

No. 11. Youth, aged 16. Sad case; Londoner. Works at odd jobs and matches selling. Has taken 3d. to-day, *i.e.*, net profit 1½d. Has five boxes still. Has slept here every night for a month. Before that slept in Covent Garden Market or on doorsteps. Been sleeping out six months, since he left Feltham Industrial School. Was sent there for playing truant. Has had one bit of bread to-day; yesterday had only some gooseberries and cherries, *i.e.*, bad ones that had been thrown away. Mother is alive. She "chucked him out" when he returned home on leaving Feltham because he could'nt find her money for drink.

No. 12. Old man, age 67. Seems to take rather a humorous view of the position. Kind of Mark Tapley. Says he can't say he does like it, but then he *must* like it! Ha, ha! Is a slater by trade. Been out of work some time; younger men naturally get the work. Gets a bit of bricklaying sometimes; can turn his hand to anything. Goes miles and gets nothing. Earned one and twopence this week at holding horses. Finds it hard, certainly. Used to care once, and get down-hearted, but that's no good; don't trouble now. Had a bit of bread and butter and cup of coffee to-day. Health is awful bad, not half the size he was; exposure and want of food is the cause; got wet last night, and is very stiff in consequence. Has been walking about since it was light, that is 3 a.m. Was so cold and wet and weak, scarcely knew what to do. Walked to Hyde Park, and got a little sleep there on a dry seat as soon as the park opened.

These are fairly typical cases of the men who are now wandering homeless through the streets. That is the way in which the nomads of civilization are constantly being recruited from above.

Such are the stories gathered at random one Midsummer night this year under the shade of the plane trees of the Embankment. A month later, when one of my staff took the census of the sleepers out of doors along the line of the Thames from Blackfriars to Westminster, he found three hundred and sixty-eight persons sleeping in the open air. Of these, two hundred and seventy were on the Embankment proper, and ninety-eight in and about Covent Garden Market, while the recesses of Waterloo and Blackfriars Bridges were full of human misery.

This, be it remembered, was not during a season of bad trade. The revival of business has been attested on all hands, notably by the barometer of strong drink. England is prosperous enough to drink rum in quantities which appall the Chancellor of the Exchequer, but she is not prosperous enough to provide other shelter than the midnight sky for these poor outcasts on the Embankment.

To very many even of those who live in London it may be news that there are so many hundreds who sleep out of doors every night. There are comparatively few people stirring after midnight, and when we are snugly tucked into our own beds we are apt to forget the multitude outside in the rain and the storm who are shivering the long hours through on the hard stone seats in the open or under the arches of the railway. These homeless, hungry people are, however, there, but being broken-spirited folk for the most part they seldom make their voices audible in the ears of their neighbours. Now and again, however, a harsh cry from the depths is heard for a moment, jarring rudely upon the ear and then all is still. The inarticulate classes speak as seldom as Balaam's ass. But they sometimes find a voice. Here for instance is one such case which impressed me much. It was reported in one of the Liverpool papers some time back. The speaker was haranguing a small knot of twenty or thirty men :—

"My lads," he commenced, with one hand in the breast of his ragged vest, and the other, as usual, plucking nervously at his beard, "This kind o' work can't last for ever." (Deep and earnest exclamations, "It can't! It shan't") "Well, boys," continued the speaker, "Somebody'll have to find a road out o' this. What we want is work, not work'us bounty, though the parish has been busy enough amongst us lately, God knows! What we want is honest work. (Hear, hear.) Now, what I propose is that each of you gets fifty mates to join you; that'll make about 1,200 starving chaps—" "And then?" asked several very gaunt and hungry-looking men excitedly. " Why, then," continued the leader. " Why, then," interrupted a cadaverous-looking man from the farther and darkest end of the cellar, " of course we'll make a——London job of it, eh ? " " No, no," hastily interposed my friend, and holding up his hands deprecatingly, " we'll go peaceably about it chaps ; we'll go in a body to the Town Hall, and show our poverty, and ask for work. We'll take the women and children with us too." ("Too ragged ! Too starved ! They can't walk it !") " The women's rags is no disgrace, the staggerin' children 'll show what we come to. Let's go a thousand strong, and ask for work and bread ! "

Three years ago, in London, there were some such processions. Church parades to the Abbey and St. Paul's, bivouacs in Trafalgar Square, etc. But Lazarus showed his rags and his sores too conspicuously for the convenience of Dives, and was summarily dealt with in the name of law and order. But as we have Lord Mayor's Days, when all the well-fed fur-clad City Fathers go in State Coaches through the town, why should we not have a Lazarus Day, in which the starving Out-of-Works, and the sweated half-starved " in-works " of London should crawl in their tattered raggedness, with their gaunt, hungry faces, and emaciated wives and children, a Procession of Despair through the main thoroughfares, past the massive houses and princely palaces of luxurious London ?

For these men are gradually, but surely, being sucked down into the quicksand of modern life. They stretch out their grimy hands to us in vain appeal, not for charity, but for work.

Work, work ! it is always work that they ask. The Divine curse is to them the most blessed of benedictions. " In the sweat of thy brow thou shalt eat thy bread," but alas for these forlorn sons of Adam, they fail to find the bread to eat, for Society has no work for them to do. They have not even leave to sweat. As well as discussing how these poor wanderers should in the second Adam "all be made alive," ought we not to put forth some effort to effect their restoration to that share in the heritage of labour which is theirs by right of descent from the first Adam ?

CHAPTER IV.

THE OUT-OF-WORKS.

There is hardly any more pathetic figure than that of the strong, able worker crying plaintively in the midst of our palaces and churches, not for charity, but for work, asking only to be allowed the privilege of perpetual hard labour, that thereby he may earn wherewith to fill his empty belly and silence the cry of his children for food. Crying for it and not getting it, seeking for labour as lost treasure and finding it not, until at last, all spirit and vigour worn out in the weary quest, the once willing worker becomes a broken-down drudge, sodden with wretchedness and despairing of all help in this world or in that which is to come. Our organisation of industry certainly leaves much to be desired. A problem which even slave owners have solved ought not to be abandoned as insoluble by the Christian civilisation of the Nineteenth Century.

I have already given a few life stories taken down from the lips of those who were found homeless on the Embankment which suggest somewhat of the hardships and the misery of the fruitless search for work. But what a volume of dull, squalid horror—a horror of great darkness gradually obscuring all the light of day from the life of the sufferer—might be written from the simple prosaic experiences of the ragged fellows whom you meet every day in the street. These men, whose labour is their only capital, are allowed, nay compelled, to waste day after day by the want of any means of employment, and then when they have seen days and weeks roll by during which their capital has been wasted by pounds and pounds, they are lectured for not saving the pence. When a rich man cannot employ his capital he puts it out at interest, but the bank for the labour capital of the poor man has yet to be invented. Yet it might be worth while inventing one. A man's labour is not only his capital, but his life. When it passes it returns never more. To utilise it, to prevent its wasteful squandering, to enable the poor man to bank it up for use hereafter, this surely is one of the most urgent tasks before civilisation.

Of all heart-breaking toil the hunt for work is surely the worst. Yet at any moment let a workman lose his present situation, and he is compelled to begin anew the dreary round of fruitless calls. Here is the story of one among thousands of the nomads, taken down from his own lips, of one who was driven by sheer hunger into crime.

A bright Spring morning found me landed from a western colony. Fourteen years had passed since I embarked from the same spot. They were fourteen years, as far as results were concerned, of non-success, and here I was again in my own land, a stranger, with a new career to carve for myself and the battle of life to fight over again.

My first thought was work. Never before had I felt more eager for a down-right good chance to win my way by honest toil; but where was I to find work? With firm determination I started in search. One day passed without success, and another, and another, but the thought cheered me, "Better luck to-morrow." It has been said, "Hope springs eternal in the human breast." In my case it was to be severely tested. Days soon ran into weeks, and still I was on the trail patiently and hopefully. Courtesy and politeness so often met me in my enquiries for employment that I often wished they would kick me out, and so vary the monotony of the sickly veneer of consideration that so thinly overlaid the indifference and the absolute unconcern they had to my needs. A few cut up rough and said, "No; we don't want you." "Please don't trouble us again (this after the second visit). We have no vacancy; and if we had, we have plenty of people on hand to fill it."

Who can express the feeling that comes over one when the fact begins to dawn that the search for work is a failure? All my hopes and prospects seemed to have turned out false. Helplessness, I had often heard of it, had often talked about it, thought I knew all about it. Yes! in others, but now I began to understand it for myself. Gradually my personal appearance faded. My once faultless linen became unkempt and unclean. Down further and further went the heels of my shoes, and I drifted into that distressing condition, "shabby gentility." If the odds were against me before, how much more so now, seeing that I was too shabby even to command attention, much less a reply to my enquiry for work.

Hunger now began to do its work, and I drifted to the dock gates, but what chance had I among the hungry giants there? And so down the stream I drifted until "Grim Want" brought me to the last shilling, the last lodging, and the last meal. What shall I do? Where shall I go? I tried to think. Must

C

I starve? Surely there must be some door still open for honest willing endeavour, but where? What can I do? "Drink," said the Tempter; but to drink to drunkenness needs cash, and oblivion by liquor demands an equivalent in the currency.

Starve or steal. "You must do one or the other," said the Tempter. But I recoiled from being a Thief. "Why be so particular?" says the Tempter again. "You are down now, who will trouble about you? Why trouble about yourself? The choice is between starving and stealing." And I struggled until hunger stole my judgment, and then I became a Thief.

No one can pretend that it was an idle fear of death by starvation which drove this poor fellow to steal. Deaths from actual hunger are more common than is generally supposed. Last year, a man, whose name was never known, was walking through St. James's Park, when three of our Shelter men saw him suddenly stumble and fall. They thought he was drunk, but found he had fainted. They carried him to the bridge and gave him to the police. They took him to St. George's Hospital, where he died. It appeared that he had, according to his own tale, walked up from Liverpool, and had been without food for five days. The doctor, however, said he had gone longer than that. The jury returned a verdict of "Death from Starvation."

Without food for five days or longer! Who that has experienced the sinking sensation that is felt when even a single meal has been sacrificed may form some idea of what kind of slow torture killed that man!

In 1888 the average daily number of unemployed in London was estimated by the Mansion House Committee at 20,000. This vast reservoir of unemployed labour is the bane of all efforts to raise the scale of living, to improve the condition of labour. Men hungering to death for lack of opportunity to earn a crust are the materials from which "blacklegs" are made, by whose aid the labourer is constantly defeated in his attempts to improve his condition.

This is the problem that underlies all questions of Trades Unionism, and all Schemes for the Improvement of the Condition of the Industrial Army. To rear any stable edifice that will not perish when the first storm rises and the first hurricane blows, it must be built not upon sand, but upon a rock. And the worst of all existing Schemes for social betterment by organisation of the skilled workers and the like is that they are founded, not upon "rock," nor even upon "sand," but upon the bottomless bog of the stratum of the Workless. It is here where we must begin. The regimentation of industrial workers who have got regular work is not so very difficult. That can be done, and is being done, by themselves. The problem that we have to face is the regimentation, the organisation, of those who have not got work, or who have only irregular work, and who from sheer pressure of absolute starvation are driven irresistibly into cut-throat competition with their better employed brothers and sisters. Skin for skin, all that a man hath, will he give for his life ; much more, then, will those who experimentally know not God give all that they might hope hereafter to have—in this world or in the world to come.

There is no gainsaying the immensity of the problem. It is appalling enough to make us despair. But those who do not put their trust in man alone, but in One who is Almighty, have no right to despair. To despair is to lose faith ; to despair is to forget God. Without God we can do nothing in this frightful chaos of human misery. But with God we can do all things, and in the faith that He has made in His image all the children of men we face even this hideous wreckage of humanity with a cheerful confidence that if we are but faithful to our own high calling He will not fail to open up a way of deliverance.

I have nothing to say against those who are endeavouring to open up a way of escape without any consciousness of God's help. For them I feel only sympathy and compassion. In so far as they are endeavouring to give bread to the hungry, clothing to the naked, and above all, work to the workless, they are to that extent endeavouring to do the will of our Father which is in Heaven, and woe be unto all those who say them nay ! But to be orphaned of all sense of the Fatherhood of God is surely not a secret source of strength. It is in most cases—it would be in my own—the secret of paralysis. If I did not feel my Father's hand in the darkness, and hear His voice in the silence of the night watches bidding me put my hand to this thing, I would shrink back dismayed ;—but as it is I dare not.

How many are there who have made similar attempts and have failed, and we have heard of them no more ! Yet none of them proposed to deal with more than the mere fringe of the evil which, God helping me, I will try to face in all its immensity. Most Schemes that are put forward for the Improvement of the Circumstances of the People are either avowedly or actually limited to those whose condition least needs amelioration. The Utopians, the economists, and most of the philanthropists propound remedies, which, if adopted to-morrow, would only affect the aristo-cracy of the miserable. It is the thrifty, the industrious, the sober, the thoughtful who can take advantage of these plans. But the thrifty, the industrious, the sober, and the thoughtful are already very well able for the most part to take care of themselves. No one will ever make even a visible dint on the Morass of Squalor who does rot deal with the improvident, the lazy, the vicious, and the criminal. The Scheme of Social Salvation is not worth discussion which is not as wide as the Scheme of Eternal Salvation set forth in the Gospel. The Glad Tidings must be to every creature, not merely to an elect few who are to be saved while the mass of their fellows are predestined to a temporal damnation. We have had this doctrine of an inhuman cast-iron pseudo-political economy too long enthroned amongst us. It is now time to fling down the false idol, and proclaim a Temporal Salvation as full, free, and universal, and with no other limitations than the "Whosoever will," of the Gospel.

To attempt to save the Lost, we must accept no limitations to human brotherhood. If the Scheme which I set forth in these and the following pages is not applicable to the Thief, the Harlot, the Drunkard, and the Sluggard, it may as well be dismissed without ceremony. As Christ came to call not the saints but sinners to repentance, so the New Message of Temporal Salvation, of salvation from pinching poverty, from rags and misery, must be offered to all. They may reject it, of course. But we who call ourselves by the name of Christ are not worthy to profess to be His disciples until we have set an open door before the least and worst of these who are now apparently imprisoned for life in a horrible dungeon of misery and despair. The responsibility for its rejection must be theirs, not ours. We all know the prayer 'Give me neither poverty nor riches, feed me with food convenient for me "—and for every child of man on this planet, thank God the prayer of Agur, the son of Jakeh, may be fulfilled.

At present how far it is from being realised may be seen by anyone who will take the trouble to go down to the docks and see the struggle for work. Here is a sketch of what was found there this summer :—

London Docks, 7.25 a.m. The three pairs of huge wooden doors are closed. Leaning against them, and standing about, there are perhaps a couple of hundred men. The public house opposite is full, doing a heavy trade. All along the road are groups of men, and from each direction a steady stream increases the crowd at the gate.

7.30. Doors open, there is a general rush to the interior. Everybody marches about a hundred yards along to the iron barrier—a temporary chain affair, guarded by the dock police. Those men who have previously (*i.e.*, night before) been engaged, show their ticket and pass through, about six hundred. The rest—some five hundred—stand behind the barrier, patiently waiting the chance of a job, but *less than twenty* of these get engaged. They are taken on by a foreman who appears next the barrier and proceeds to pick his men. No sooner is the foreman seen, than there is a wild rush to the spot and a sharp, mad fight to "catch his eye." The men picked out, pass the barrier, and the excitement dies away until another lot of men are wanted.

They wait until eight o'clock strikes, which is the signal to withdraw. The barrier is taken down and all those hundreds of men, wearily disperse to "find a job." Five hundred applicants, twenty acceptances! No wonder one tired-out looking individual ejaculates, "Oh dear, Oh dear! Whatever shall I do?" A few hang about until mid-day on the slender chance of getting taken on then for half a day.

Ask the men and they will tell you something like the following story, which gives the simple experiences of a dock labourer.

R. P. said :—"I was in regular work at the South West India Docks before the strike. We got 5d. an hour. Start work 8 a.m. summer and 9 a.m. winter. Often there would be five hundred go, and only twenty get taken on (that is besides those engaged the night previous.) The foreman stood in his box, and called out the men he wanted. He would know quite five hundred by name. It was a regular fight to get work, I have known nine hundred to be taken on, but there's always hundreds turned away. You see they get to know when ships come in, and when they're consequently likely to be wanted, and turn up then in greater numbers. I would earn 30s. a week sometimes and then perhaps nothing for a fortnight. That's what makes it so hard. You get nothing to eat for a week scarcely, and then when you get taken on, you are so weak that you can't do it properly. I've stood in the crowd at the gate and had to go away without work, hundreds of times. Still I should go at it again if I could. I got tired of the little work and went away into the country to get work on a farm, but couldn't get it, so I'm without the 10s. that it costs to join the Dockers' Union. I'm going to the country again in a day or two to try again. Expect to get 3s. a day perhaps. Shall come back to the docks again. There *is* a chance of getting regular dock work, and that is, to lounge about the pubs, where the foremen go, and treat them. Then they will very likely take you on next day."

R. P. was a non-Unionist. Henry F. is a Unionist. His history is much the same.

"1 worked at St. Katherine's Docks five months ago. You have to get to the gates at 6 o'clock for the first call. There's generally about 400 waiting. They will take on one to two hundred. Then at 7 o'clock there's a second call. Another 400 will have gathered by then, and another hundred or so will be taken on. Also there will probably be calls at nine and one o'clock. About the same number turn up but there's no work for many hundreds of them. I was a Union man. That means 10s. a week sick pay, or 8s. a week for slight accidents; also some other advantages. The Docks won't take men on now unless they are Unionists. The point is that there's too many men. I would often be out of work a fortnight to three weeks at a time. Once earned £3 in a week, working day and night, but then had a fortnight out directly after. Especially when there don't happen to be any ships in for a few days, which means, of course, nothing to unload. That's the time; there's plenty of men almost starving then. They have no trade to go to, or can get no work at it, and they swoop do to the docks for work, when they had much better stay away."

But it is not only at the dock-gates that you come upon these unfortunates who spend their lives in the vain hunt for work. Here is the story of another man whose case has only too many parallels.

C. is a fine built man, standing nearly six feet. He has been in the Royal Artillery for eight years and held very good situations whilst in it. It seems that he was thrifty and consequently steady. He bought his discharge, and being an excellent cook opened a refreshment house, but at the end of five months he was compelled to close his shop on account of slackness in trade, which was brought about by the closing of a large factory in the locality.

After having worked in Scotland and Newcastle-on-Tyne for a few years, and through ill health having to give up his situation, he came to London with the hope that he might get something to do in his native town. He has had no regular employment for the past eight months. His wife and family are in a state of destitution, and he remarked, "We only had 1 lb. of bread between us yesterday." He is six weeks in arrears of rent, and is afraid that he will be ejected. The furniture which is in his home is not worth 3s. and the clothes of each member of his family are in a tattered state and hardly fit for the rag bag. He assured us he had tried everywhere to get employment and would be willing to take anything. His characters are very good indeed.

Now, it may seem a preposterous dream that any arrangement can be devised by which it may be possible, under all circumstances, to provide food, clothes, and shelter for all these Out-of-Works without any loss of self respect; but I am convinced that it can be done, providing only that they are willing to Work, and, God helping me, if the means are forthcoming, I mean to try to do it; how, and where, and when, I will explain in subsequent chapters.

All that I need say here is, that so long as a man or woman is willing to submit to the discipline indispensable in every campaign against any formidable foe, there appears to me nothing impossible about this ideal; and the great element of hope before us is that the majority are, beyond all gainsaying, eager for work. Most of them now do more exhausting work in seeking for employment than the regular toilers do in their workshops, and do it, too, under the darkness of hope deferred which maketh the heart sick.

CHAPTER V.

ON THE VERGE OF THE ABYSS.

There is, unfortunately, no need for me to attempt to set out, how-ever imperfectly, any statement of the evil case of the sufferers whom we wish to help. For years past the Press has been filled with echoes of the "Bitter Cry of Outcast London," with pictures of "Horrible Glasgow," and the like. We have had several volumes describing "How the Poor Live," and I may therefore assume that all my readers are more or less cognizant of the main outlines of "Darkest England." My slum officers are living in the midst of it; their reports are before me, and one day I may publish some more detailed account of the actual facts of the social condition of the Sunken Millions. But not now. All that must be taken as read. I only glance at the subject in order to bring into clear relief the salient points of our new Enterprise.

I have spoken of the houseless poor. Each of these represents a point in the scale of human suffering below that of those who have still contrived to keep a shelter over their heads. A home is a home, be it ever so low; and the desperate tenacity with which the poor will cling to the last wretched semblance of one is very touch-ing. There are vile dens, fever-haunted and stenchful crowded courts, where the return of summer is dreaded because it means the unloosing of myriads of vermin which render night unbearable, which, nevertheless, are regarded at this moment as havens of rest by their hard-working occupants. They can scarcely be said to be furnished. A chair, a mattress, and a few miserable sticks constitute all the furniture of the single room in which they have to sleep, and breed, and die; but they cling to it as a drowning man to a half-submerged raft. Every week they contrive by pinch-ing and scheming to raise the rent, for with them it is pay or go; and they struggle to meet the collector as the sailor nerves himself to avoid being sucked under by the foaming wave. If at any time work fails or sickness comes they are liable to drop helplessly into the ranks of the homeless. It is bad for a single man to have to confront the struggle for life in the streets and Casual Wards. But how much more terrible must it be for the married man with his wife and children to be turned out into the streets. So long as the family has a lair into which it can creep at night, he keeps his footing; but when he loses that solitary foothold then arrives the time if there be such athing as Christian compassion, for the helping hand to be held cut to save him from the vortex that sucks him downward —ay, downward to the hopeless under-strata of crime and despair.

"The heart knoweth its own bitterness and the stranger inter-meddleth not therewith." But now and then out of the depths there sounds a bitter wail as of some strong swimmer in his agony as he is drawn under by the current. A short time ago a respectable man, a chemist in Holloway, fifty years of age, driven hard to the wall, tried to end it all by cutting his throat. His wife also cut her throat, and at the same time they gave strychnine to their only child. The effort failed, and they were placed on trial for attempted murder. In the Court a letter was read which the poor wretch had written before attempting his life :—

MY DEAREST GEORGE,—Twelve months have I now passed of a most miserable and struggling existence, and I really cannot stand it any more. I am completely worn out, and relations who could assist me won't do any more, for such was uncle's last intimation. Never mind; he can't take his money and comfort with him, and in all probability will find himself in the same boat as myself. He never enquires whether I am starving or not. £3—a mere flea-bite to him—would have put us straight, and with his security and good interest might have obtained me a good situation long ago. I can face poverty and degradation no longer, and would sooner die than go to the workhouse, whatever may be the awful consequences of the steps we have taken. We have, God forgive us, taken our darling Arty with us out of pure love and affection, so that the darling should never be cuffed about, or reminded or taunted with his heart-broken parents' crime. My poor wife has done her best at needle-work, washing, house-minding, &c., in fact, anything and everything that would bring in a shilling; but it would only keep us in semi-starvation. I have now done six weeks' travelling from morning till night, and not received one farthing for it. If that is not enough to drive you mad—wickedly mad—I don't know what is. No bright prospect anywhere; no ray of hope.

May God Almighty forgive us for this heinous sin, and have mercy on our sinful souls, is the prayer of your miserable, broken-hearted, but loving brother, Arthur. We have now done everything that we can possibly think of to avert this wicked proceeding, but can discover no ray of hope. Fervent prayer has availed us nothing; our lot is cast, and we must abide by it. It must be God's will or He would have ordained it differently. Dearest Georgy, I am exceedingly sorry to leave you all, but I am mad—thoroughly mad. You, dear, must try and forget us, and, if possible, forgive us; for I do not consider it our own fault we have not succeeded. If you could get £3 for our bed it will pay our rent, and our scanty furniture may fetch enough to bury us in a cheap way. Don't grieve over us or follow us, for we shall not be worthy of such respect. Our clergyman has never called on us or given us the least consolation, though I called on him a month ago. He is paid to preach, and there he considers his responsibility ends, the rich excepted. We have only yourself and a very few others who care one pin what becomes of us, but you must try and forgive us, is the last fervent prayer of your devotedly fond and affectionate but broken-hearted and persecuted brother.   (Signed) R. A. O——.

That is an authentic human document—a transcript from the life of one among thousands who go down inarticulate into the depths. They die and make no sign, or, worse still, they continue to exist, carrying about with them, year after year, the bitter ashes of a life from which the furnace of misfortune has burnt away all joy, and hope, and strength. Who is there who has not been confronted by many despairing ones, who come, as Richard O—— went, to the clergyman, crying for help, and how seldom have we been able to give it them? It is unjust, no doubt, for them to blame the clergy and the comfortable well-to-do—for what can they do but preach and offer good advice? To assist all the Richard O——s' by direct financial advance would drag even Rothschild into the gutter. And what else can be done? Yet something else must be done if Christianity is not to be a mockery to perishing men.

Here is another case, a very common case, which illustrates how the Army of Despair is recruited.

Mr. T., Margaret Place, Gascoign Place, Bethnal Green, is a bootmaker by trade. Is a good hand, and has earned three shillings and sixpence to four shillings and sixpence a day. He was taken ill last Christmas, and went to the London Hospital; was there three months. A week after he had gone Mrs. T. had rheumatic fever, and was taken to Bethnal Green Infirmary, where she remained about three months. Directly after they had been taken ill, their furniture was seized for the three weeks' rent which was owing. Consequently, on becoming convalescent, they were homeless. They came out about the same time. He went out to a lodging-house for a night or two, until she came out. He then had twopence, and she had sixpence, which a nurse had given her.   They went to a lodging-house together, but the society there was dreadful.   Next day he had a day's work, and got two shillings and sixpence, and on the strength of this they took a furnished room at tenpence per day (payable nightly).   His work lasted a few weeks, when he was again taken ill, lost his job, and spent all their money. Pawned a shirt and apron for a shilling; spent that, too.   At last pawned their tools for three shillings, which got them a few days' food and lodging.   He is now minus tools and cannot work at his own job, and does anything he can. Spent their last twopence on a pen'orth each of tea and sugar.   In two days they had a slice of bread and butter each, that's all.   They are both very weak through want of food.

" Let things alone," the laws of supply and demand, and all the rest of the excuses by which those who stand on firm ground salve their consciences when they leave their brother to sink, how do they look when we apply them to the actual loss of life at sea ? Does "Let things alone" man the lifeboat ? Will the inexorable laws of political economy save the shipwrecked sailor from the boiling surf ? They often enough are responsible for his disaster.   Coffin ships are a direct result of the wretched policy of non-interference with the legitimate operations of commerce, but no desire to make it pay created the National Lifeboat Institution, no law of supply and demand actuates the volunteers who risk their lives to bring the shipwrecked to shore.

What we have to do is to apply the same principle to society. We want a Social Lifeboat Institution, a Social Lifeboat Brigade, to snatch from the abyss those who, if left to themselves, will perish as miserably as the crew of a ship that founders in mid-ocean.

The moment that we take in hand this work we shall be compelled to turn our attention seriously to the question whether prevention is not better than cure.   It is easier and cheaper, and in every way better, to prevent the loss of home than to have to re-create that home.   It is better to keep a man out of the mire than to let him fall in first and then risk the chance of plucking him out.   Any Scheme, therefore, that attempts to deal with the reclamation of the lost must tend to develop into an endless variety of ameliorative measures, of some of which I shall have somewhat to say hereafter.   I only mention the subject here in order that no one may say I am blind to the necessity of going further and adopting wider plans of operation than those which I put forward in this book.   The renovation of our Social System is a work so vast that no one of us, nor all of us put together, can define all the measures that will have to be taken before we attain even the Cab-Horse Ideal of existence for our children and children's children. All that we can do is to attack, in a serious, practical spirit the worst and most pressing evils, knowing that if we do our duty we obey the voice of God. He is the Captain of our Salvation. If we but follow where He leads we shall not want for marching orders, nor need we imagine that He will narrow the field of operations.

I am labouring under no delusions as to the possibility of inaugurating the Millennium by any social specific. In the struggle of life the weakest will go to the wall, and there are so many weak. The fittest, in tooth and claw, will survive. All that we can do is to soften the lot of the unfit and make their suffering less horrible than it is at present. No amount of assistance will give a jellyfish a backbone. No outside propping will make some men stand erect. All material help from without is useful only in so far as it develops moral strength within. And some men seem to have lost even the very faculty of self-help. There is an immense lack of common sense and of vital energy on the part of multitudes.

It is against Stupidity in every shape and form that we have to wage our eternal battle. But how can we wonder at the want of sense on the part of those who have had no advantages, when we see such plentiful absence of that commodity on the part of those who have had all the advantages?

How can we marvel if, after leaving generation after generation to grow up uneducated and underfed, there should be developed a heredity of incapacity, and that thousands of dull-witted people should be born into the world, disinherited before their birth of their share in the average intelligence of mankind?

Besides those who are thus hereditarily wanting in the qualities necessary to enable them to hold their own, there are the weak, the disabled, the aged, and the unskilled; worse than all, there is the want of character. Those who have the best of reputation, if they lose their foothold on the ladder, find it difficult enough to regain their place. What, then, can men and women who have no character do? When a master has the choice of a hundred honest men, is it reasonable to expect that he will select a poor fellow with tarnished reputation?

All this is true, and it is one of the things that makes the problem almost insoluble. And insoluble it is, I am absolutely convinced, unless it is possible to bring new moral life into the soul of these people. This should be the first object of every social reformer, whose work will only last if it is built on the solid foundation of a new birth, to cry "You must be born again."

To get a man soundly saved it is not enough to put on him a pair of new breeches, to give him regular work, or even to give him a University education. These things are all outside a man, and if the inside remains unchanged you have wasted your labour. You must in some way or other graft upon the man's nature a new nature, which has in it the element of the Divine. All that I propose in this book is governed by that principle.

The difference between the method which seeks to regenerate the man by ameliorating his circumstances and that which ameliorates his circumstances in order to get at the regeneration of his heart, is the difference between the method of the gardener who grafts a Ribstone Pippin on a crab-apple tree and one who merely ties apples with string upon the branches of the crab. To change the nature of the individual, to get at the heart, to save his soul is the only real, lasting method of doing him any good. In many modern schemes of social regeneration it is forgotten that "it takes a soul to move a body, e'en to a cleaner sty," and at the risk of being mis-understood and misrepresented, I must assert in the most un-qualified way that it is primarily and mainly for the sake of saving the soul that I seek the salvation of the body

But what is the use of preaching the Gospel to men whose whole attention is concentrated upon a mad, desperate struggle to keep themselves alive ? You might as well give a tract to a shipwrecked sailor who is battling with the surf which has drowned his comrades and threatens to drown him. He will not listen to you. Nay, he cannot hear you any more than a man whose head is under water can listen to a sermon. The first thing to do is to get him at least a footing on firm ground, and to give him room to live. Then you may have a chance. At present you have none. And you will have all the better opportunity to find a way to his heart, if he comes to know that it was you who pulled him out of the horrible pit and the miry clay in which he was sinking to perdition.

CHAPTER VI.

THE VICIOUS.

There are many vices and seven deadly sins. But of late years many of the seven have contrived to pass themselves off as virtues. Avarice, for instance ; and Pride, when re-baptised thrift and self-respect, have become the guardian angels of Christian civilisation ; and as for Envy, it is the corner-stone upon which much of our competitive system is founded. There are still two vices which are fortunate, or unfortunate, enough to remain undisguised, not even concealing from themselves the fact that they are vices and not virtues. One is drunkenness ; the other fornication. The viciousness of these vices is so little disguised, even from those who habitually practise them, that there will be a protest against merely describing one of them by the right Biblical name. Why not say prostitution ? For this reason : prostitution is a word applied to only one half of the vice, and that the most pitiable. Fornication hits both sinners alike. Prostitution applies only to the woman.

When, however, we cease to regard this vice from the point of view of morality and religion, and look at it solely as a factor in the social problem, the word prostitution is less objectionable. For the social burden of this vice is borne almost entirely by women. The male sinner does not, by the mere fact of his sin, find himself in a worse position in obtaining employment, in finding a home, or even in securing a wife. His wrong-doing only hits him in his purse, or, perhaps, in his health. His incontinence, excepting so far as it relates to the woman whose degradation it necessitates, does not add to the number of those for whom society has to provide. It is an immense addition to the infamy of this vice in man that its consequences have to be borne almost exclusively by woman.

The difficulty of dealing with drunkards and harlots is almost insurmountable. Were it not that I utterly repudiate as a fundamental denial of the essential principle of the Chstian religion the popular pseudo-scientific doctrine that any man or woman is past saving by the grace of God and the power of the Holy Spirit, I would sometimes be disposed to despair when contemplating these victims of the Devil. The doctrine of Heredity and the suggestion of Irresponsibility come perilously near re-establishing, on scientific bases, the awful dogma of Reprobation which has cast so terrible a shadow over the Christian Church. For thousands upon thousands of these poor wretches are, as Bishop South truly said, "not so much born into this world as damned into it." The bastard of a harlot, born in a brothel, suckled on gin, and familiar from earliest infancy with all the bestialities of debauch, violated before she is twelve, and driven out into the streets by her mother a year or two later, what chance is there for such a girl in this world—I say nothing about the next? Yet such a case is not exceptional. There are many such differing in detail, but in essentials the same. And with boys it is almost as bad. There are thousands who were begotten when both parents were besotted with drink, whose mothers saturated themselves with alcohol every day of their pregnancy, who may be said to have sucked in a taste for strong drink with their mothers' milk, and who were surrounded from childhood with opportunities and incitements to drink. How can we marvel that the constitution thus disposed to intemperance finds the stimulus of drink indispensable? Even if they make a stand against it, the increasing pressure of exhaustion and of scanty food drives them back to the cup. Of these poor wretches, born slaves of the bottle, predestined to drunkenness from their mother's womb, there are—who can say how many? Yet they are all men; all with what the Russian peasants call "a spark of God" in them, which can never be wholly obscured and destroyed while life exists, and if any social scheme is to be comprehensive and practical it must deal with these men. It must provide for the drunkard and the harlot as it provides for the improvident and the out-of-work. But who is sufficient for these things?

I will take the question of the drunkard, for the drink difficulty lies at the root of everything. Nine-tenths of our poverty, squalor, vice, and crime spring from this poisonous tap-root. Many of our social evils, which overshadow the land like so many upas trees, would dwindle away and die if they were not constantly watered with strong drink. There is universal agreement on that point; in fact, the

agreement as to the evils of intemperance is almost as universal as
the conviction that politicians will do nothing practical to interfere
with them. In Ireland, Mr. Justice Fitzgerald says that intemperance
leads to nineteen-twentieths of the crime in that country, but no one
proposes a Coercion Act to deal with that evil. In England,
the judges all say the same thing. Of course it is a mistake
to assume that a murder, for instance, would never be committed by
sober men, because murderers in most cases prime themselves for
their deadly work by a glass of Dutch courage. But the facility of
securing a reinforcement of passion undoubtedly tends to render
always dangerous, and sometimes irresistible, the temptation to violate
the laws of God and man.

Mere lectures against the evil habit are, however, of no avail.
We have to recognise, that the gin-palace, like many other evils,
although a poisonous, is still a natural outgrowth of our social conditions. The tap-room in many cases is the poor man's only parlour.
Many a man takes to beer, not from the love of beer, but from a
natural craving for the light, warmth, company, and comfort which is
thrown in along with the beer, and which he cannot get excepting by
buying beer. Reformers will never get rid of the drink shop until
they can outbid it in the subsidiary attractions which it offers to its
customers. Then, again, let us never forget that the temptation to
drink is strongest when want is sharpest and misery the most acute.
A well-fed man is not driven to drink by the craving that torments
the hungry ; and the comfortable do not crave for the boon of forgetfulness. Gin is the only Lethe of the miserable. The foul and
poisoned air of the dens in which thousands live predisposes to a
longing for stimulant. Fresh air, with its oxygen and its ozone,
being lacking, a man supplies the want with spirit. After a time the
longing for drink becomes a mania. Life seems as insupportable without alcohol as without food. It is a disease often inherited, always developed by indulgence, but as clearly a disease as ophthalmia or stone.

All this should predispose us to charity and sympathy. While
recognising that the primary responsibility must always rest upon
the individual, we may fairly insist that society, which, by its habits,
its customs, and its laws, has greased the slope down which these
poor creatures slide to perdition, shall seriously take in hand their
salvation.

How many are there who are, more or less, under the dominion
of strong drink ? Statistics abound, but they seldom tell us what

we want to know.   We know how many public-houses there are in the land, and how many arrests for drunkenness the police make in a year; but beyond that we know little.   Everyone knows that for one man who is arrested for drunkenness there are at least ten— and often twenty—who go home intoxicated.   In London, for instance, there are 14,000 drink shops, and every year 20,000 persons are arrested for drunkenness.   But who can for a moment believe that there are only 20,000, more or less, habitual drunkards in London ?   By habitual drunkard I do not mean one who is always drunk, but one who is so much under the dominion of the evil habit that he cannot be depended upon not to get drunk whenever the opportunity offers.

In the United Kingdom there are 190,000 public-houses, and every year there are 200,000 arrests for drunkenness.   Of course, several of these arrests refer to the same person, who is locked up again and again.   Were this not so, if we allowed six drunkards to each house as an average, or five habitual drunkards for one arrested for drunkenness, we should arrive at a total of a million adults who are more or less prisoners of the publican—as a matter of fact, Wm. Hoyle gives 1 in 12 of the adult population.   This may be an excessive estimate, but, if we take half of a million, we shall not be accused of exaggeration.   Of these some are in the last stage of confirmed dipsomania ; others are but over the verge; but the procession tends ever downwards.

The loss which the maintenance of this huge standing army of a half of a million of men who are more or less always besotted men whose intemperance impairs their working power, consumes their earnings, and renders their homes wretched, has long been a familiar theme of the platform.   But what can be done for them ?   Total abstinence is no doubt admirable, but how are you to get them to be totally abstinent ?   When a man is drowning in mid-ocean the one thing that is needful, no doubt, is that he should plant his feet firmly on terra firma.   But how is he to get there ?   It is just what he cannot do. And so it is with the drunkards. If they are to be rescued there must be something more done for them than at present is attempted, unless, of course, we decide definitely to allow the iron laws of nature to work themselves out in their destruction.   In that case it might be more merciful to facilitate the slow workings of natural law.   There is no need of establishing a lethal chamber for drunkards like that into which the lost dogs of London are

D

driven, to die in peaceful sleep under the influence of carbonic oxide. The State would only need to go a little further than it goes at present in the way of supplying poison to the community. If, in addition to planting a flaming gin palace at each corner, free to all who enter, it were to supply free gin to all who have attained a certain recognised standard of inebriety, delirium tremens would soon reduce our drunken population to manageable proportions. I can imagine a cynical millionaire of the scientific philanthropic school making a clearance of all the drunkards in a district by the simple expedient of an unlimited allowance of alcohol. But that for us is out of the question. The problem of what to do with our half of a million drunkards remains to be solved, and few more difficult questions confront the social reformer.

The question of the harlots is, however, quite as insoluble by the ordinary methods. For these unfortunates no one who looks below the surface can fail to have the deepest sympathy. Some there are, no doubt, perhaps many, who—whether from inherited passion or from evil education—have deliberately embarked upon a life of vice, but with the majority it is not so. Even those who deliberately and of free choice adopt the profession of a prostitute, do so under the stress of temptations which few moralists seem to realise. Terrible as the fact is, there is no doubt it is a fact that there is no industrial career in which for a short time a beautiful girl can make as much money with as little trouble as the profession of a courtesan. The case recently tried at the Lewes assizes, in which the wife of an officer in the army admitted that while living as a kept mistress she had received as much as £4,000 a year, was no doubt very exceptional. Even the most successful adventuresses seldom make the income of a Cabinet Minister. But take women in professions and in businesses all round, and the number of young women who have received £500 in one year for the sale of their person is larger than the number of women of all ages who make a similar sum by honest industry. It is only the very few who draw these gilded prizes, and they only do it for a very short time. But it is the few prizes in every profession which allure the multitude, who think little of the many blanks. And speaking broadly, vice offers to every good-looking girl during the first bloom of her youth and beauty more money than she can earn by labour in any field of industry open to her sex. The penalty exacted afterwards is disease, degradation and death, but these things at first are hidden from her sight.

The profession of a prostitute is the only career in which the maximum income is paid to the newest apprentice. It is the one calling in which at the beginning the only exertion is that of self-indulgence; all the prizes are at the commencement. It is the ever-new embodiment of the old fable of the sale of the soul to the Devil. The tempter offers wealth, comfort, excitement, but in return the victim must sell her soul, nor does the other party forget to exact his due to the uttermost farthing. Human nature, however, is short-sighted. Giddy girls, chafing against the restraints of uncongenial industry, see the glittering bait continually before them. They are told that if they will but " do as others do " they will make more in a night, if they are lucky, than they can make in a week at their sewing; and who can wonder that in many cases the irrevocable step is taken before they realise that it is irrevocable, and that they have bartered away the future of their lives for the paltry chance of a year's ill-gotten gains ?

Of the severity of the punishment there can be no question. If the premium is high at the beginning, the penalty is terrible at the close. And this penalty is exacted equally from those who have deliberately said, " Evil, be thou my Good," and for those who have been decoyed, snared, trapped into the life which is a living death. When you see a girl on the street you can never say without enquiry whether she is one of the most-to-be condemned, or the most-to-be pitied of her sex. Many of them find themselves where they are because of a too trusting disposition, confidence born of innocence being often the unsuspecting ally of the procuress and seducer. Others are as much the innocent victims of crime as if they had been stabbed or maimed by the dagger of the assassin. The records of our Rescue Homes abound with life-stories, some of which we have been able to verify to the letter—which prove only too conclusively the existence of numbers of innocent victims whose entry upon this dismal life can in no way be attributed to any act of their own will. Many are orphans or the children of depraved mothers, whose one idea of a daughter is to make money out of her prostitution. Here are a few cases on our register :—

E. C., aged 18, a soldier's child, born on the sea. Her father died, and her mother, a thoroughly depraved woman, assisted to secure her daughter's prostitution.

P. S., aged 20, illegitimate child. Went to consult a doctor one time about some ailment. The doctor abused his position and took advantage of his patient, and when she complained, gave her £4 as compensation. When that was spent, having lost her character, she came on the town. We looked the doctor up, and he fled.

E. A., aged 17, was left an orphan very early in life, and adopted by her god-father, who himself was the means of her ruin at the age of 10.

A girl in her teens lived with her mother in the "Dusthole," the lowest part of Woolwich. This woman forced her out upon the streets, and profited by her prostitution up to the very night of her confinement. The mother had all the time been the receiver of the gains.

E., neither father nor mother, was taken care of by a grandmother till, at an early age, accounted old enough. Married a soldier; but shortly before the birth of her first child, found that her deceiver had a wife and family in a distant part of the country, and she was soon left friendless and alone. She sought an asylum in the Workhouse for a few weeks, after which she vainly tried to get honest employment. Failing that, and being on the very verge of starvation, she entered a lodging-house in Westminster and "did as other girls." Here our lieutenant found and persuaded her to leave and enter one of our Homes, where she soon gave abundant proof of her conversion by a thoroughly changed life. She is now a faithful and trusted servant in a clergyman's family.

A girl was some time ago discharged from a city hospital after an illness. She was homeless and friendless, an orphan, and obliged to work for her living. Walking down the street and wondering what she should do next, she met a girl, who came up to her in a most friendly fashion and speedily won her confidence.

"Discharged ill, and nowhere to go, are you?" said her new friend. "Well, come home to my mother's; she will lodge you, and we'll go to work together, when you are quite strong."

The girl consented gladly, but found herself conducted to the very lowest part of Woolwich and ushered into a brothel; there was no mother in the case. She was hoaxed, and powerless to resist. Her protestations were too late to save her, and having had her character forced from her she became hopeless, and stayed on to live the life of her false friend.

There is no need for me to go into the details of the way in which men and women, whose whole livelihood depends upon their success in disarming the suspicions of their victims and luring them to their doom, contrive to overcome the reluctance of the young girl without parents, friends, or helpers to enter their toils. What fraud fails to accomplish, a little force succeeds in effecting; and a girl who has been guilty of nothing but imprudence finds herself an outcast for life.

The very innocence of a girl tells against her. A woman of the world, once entrapped, would have all her wits about her to extricate herself from the position in which she found herself. A perfectly virtuous girl is often so overcome with shame and horror that there seems nothing in life worth struggling for. She accepts her doom without further struggle, and treads the long and torturing path-way of "the streets" to the grave.

"Judge not, that ye be not judged" is a saying that applies most appropriately of all to these unfortunates. Many of them would have escaped their evil fate had they been less innocent. They are where they are because they loved too utterly to calculate consequences, and trusted too absolutely to dare to suspect evil. And others are there because of the false education which confounds ignorance with virtue, and throws our young people into the midst of a great city, with all its excitements and all its temptations, without more preparation or warning than if they were going to live in the Garden of Eden.

Whatever sin they have committed, a terrible penalty is exacted. While the man who caused their ruin passes as a respectable member of society, to whom virtuous matrons gladly marry—if he is rich—their maiden daughters, they are crushed beneath the millstone of social excommunication.

Here let me quote from a report made to me by the head of our Rescue Homes as to the actual life of these unfortunates.

The following hundred cases are taken as they come from our Rescue Register. The statements are those of the girls themselves. They are certainly frank, and it will be noticed that only two out of the hundred allege that they took to the life out of poverty :—

CAUSE OF FALL.				CONDITION WHEN APPLYING.			
Drink...	...	...	14				
Seduction	...	...	33	Rags ...	...	...	25
Wilful choice	...	...	24	Destitution	...	...	27
Bad company	...	...	27	Decently dressed	...	48
Poverty	...	...	2				
		Total	100			Total	100

Out of these girls twenty-three have been in prison.

The girls suffer so much that the shortness of their miserable life is the only redeeming feature. Whether we look at the wretchedness of the life itself; their perpetual intoxication; the cruel treatment to which they are subjected by their task-masters and mistresses or bullies; the hopelessness, suffering and despair induced by their circumstances and surroundings; the depths of misery, degradation and poverty to which they eventually descend; or their treatment in sickness, their friendlessness and loneliness in death, it must be admitted that a more dismal lot seldom falls to the fate of a human being. I will take each of these in turn.

HEALTH.—This life induces insanity, rheumatism, consumption, and all forms of syphilis. Rheumatism and gout are the commonest of these evils. Some were quite crippled by both—young though they were. Consumption sows its seeds broadcast. The life is a hot-bed for the development of any constitutional and hereditary germs of the disease. We have found girls in Piccadilly at midnight who are continually prostrated by hæmorrhage, yet who have no other way of life open, so struggle on in this awful manner between whiles.

DRINK.—This is an inevitable part of the business. All confess that they could never lead their miserable lives if it were not for its influence.

A girl, who was educated at college, and who had a home in which was every comfort, but who, when ruined, had fallen even to the depth of Woolwich " Dusthole," exclaimed to us indignantly—" Do you think I could ever, ever do this if it weren't for the drink ? I always have to be in drink if I want to sin." No girl has ever come into our Homes *from street-life* but has been more or less a prey to drink.

CRUEL TREATMENT.—The devotion of these women to their bullies is as remarkable as the brutality of their bullies is abominable. Probably the primary cause of the fall of numberless girls of the lower class, is their great aspiration to the dignity of wifehood ;—they are never " somebody " until they are married, and will link themselves to any creature, no matter how debased, in the hope of being ultimately married by him. This consideration, in addition to their helpless condition when once character has gone, makes them suffer cruelties which they would never otherwise endure from the men with whom large numbers of them live.

One case in illustration of this is that of a girl who was once a respectable servant, the daughter of a police sergeant. She was ruined, and shame led her to leave home. At length she drifted to Woolwich, where she came across a man who persuaded her to live with him, and for a considerable length of time she kept him, although his conduct to her was brutal in the extreme.

The girl living in the next room to her has frequently heard him knock her head against the wall, and *pound* it, when he was out of temper, through her gains of prostitution being less than usual. He lavished upon her every sort of cruelty and abuse, and at length she grew so wretched, and was reduced to so dreadful a plight, that she ceased to attract. At this he became furious, and pawned all her clothing but one thin garment of rags. The week before her first confinement he kicked her black and blue from neck to knees, and she was carried to the police station in a pool of blood, but she was so loyal to the wretch that she refused to appear against him.

She was going to drown herself in desperation, when our Rescue Officers spoke to her, wrapped their own shawl around her shivering shoulders, took her home with them, and cared for her.   The baby was born dead—a tiny, shapeless mass.   This state of things is all too common.

HOPELESSNESS—SURROUNDINGS.—The state of hopelessness and despair in which these girls live continually, makes them reckless of consequences, and large numbers commit suicide who are never heard of.   A West End policeman assured us that the number of prostitute-suicides was terribly in advance of anything guessed at by the public.

DEPTHS TO WHICH THEY SINK.—There is scarcely a lower class of girls to be found than the girls of Woolwich "Dusthole"—where one of our Rescue Slum Homes is established.   The women living and following their dreadful business in this neighbourhood are so degraded that even abandoned men will refuse to accompany them home.   Soldiers are forbidden to enter the place, or to go down the street, on pain of twenty-five days' imprisonment; pickets are stationed at either end to prevent this.   The streets are much cleaner than many of the rooms we have seen.

One public house there is shut up three or four times in a day sometimes for fear of losing the licence through the terrible brawls which take place within. A policeman never goes down this street alone at night—one having died not long ago from injuries received there—but our two lasses go unharmed and loved at all hours, spending every other night always upon the streets.

The girls sink to the "Dusthole" after coming down several grades.   There is but one on record who came there with beautiful clothes, and this poor girl, when last seen by the officers, was a pauper in the workhouse infirmary in a wretched condition.

The lowest class of all is the girls who stand at the pier-head—these sell themselves literally for a bare crust of bread and sleep in the streets.

Filth and vermin abound to an extent to which no one who has not seen it can have any idea.

The "Dusthole" is only one, alas of many similar districts in this highly civilised land.

SICKNESS, FRIENDLESSNESS—DEATH.—In hospitals it is a known fact that these girls are not treated at all like other cases; they inspire disgust, and are most frequently discharged before being really cured.

Scorned by their relations, and ashamed to make their case known even to those who would help them, unable longer to struggle out on the streets to earn the bread of shame, there are girls lying in many a dark hole in this big city positively rotting away, and maintained by their old companions on the streets.

Many are totally friendless, utterly cast out and left to perish by relatives and friends.   One of this class came to us, sickened and died, and we buried her, being her only followers to the grave.

It is a sad story, but one that must not be forgotten, for these women constitute a large standing army whose numbers no one can calculate. All estimates that I have seem purely imaginary. The ordinary figure given for London is from 60,000 to 80,000. This may be true if it is meant to include all habitually unchaste women. It is a monstrous exaggeration if it is meant to apply to those who make their living solely and habitually by prostitution. These figures, however, only confuse. We shall have to deal with hundreds every month, whatever estimate we take. How utterly unprepared society is for any such systematic reformation may be seen from the fact that even now at our Homes we are unable to take in all the girls who apply. They cannot escape, even if they would, for want of funds whereby to provide them a way of release.

CHAPTER VII.

THE CRIMINALS.

One very important section of the denizens of Darkest England are the criminals and the semi-criminals. They are more or less predatory, and are at present shepherded by the police and punished by the gaoler. Their numbers cannot be ascertained with very great precision, but the following figures are taken from the prison returns of 1889 :—

The criminal classes of Great Britain, in round figures, sum up a total of no less than 90,000 persons, made up as follows :—

Convict prisons contain	11,660 persons.
Local „ „	20,883 „
Reformatories for children convicted of crime ...	1,270 „
Industrial schools for vagrant and refractory children	21,413 „
Criminal lunatics under restraint	910 „
Known thieves at large	14,747 „
Known receivers of stolen goods	1,121 „
Suspected persons	17,042 „
Total	89,046

The above does not include the great army of known prostitutes, nor the keepers and owners of brothels and disorderly houses, as to whose numbers Government is rigidly silent.

These figures are, however, misleading. They only represent the criminals actually in gaol on a given day. The average gaol population in England and Wales, excluding the convict establishments, was, in 1889, 15,119, but the total number actually sentenced and imprisoned in local prisons was 153,000, of whom 25,000 only came on first term sentences ; 76,300 of them had been convicted at least 10 times. But even if we suppose that the criminal class numbers

no more than 90,000, of whom only 35,000 persons are at large, it is
still a large enough section of humanity to compel attention. 90,000
criminals represents a wreckage whose cost to the community is very
imperfectly estimated when we add up the cost of the prisons, even
if we add to them the whole cost of the police. The police have so
many other duties besides the shepherding of criminals that it is
unfair to saddle the latter with the whole of the cost of the constabu-
lary. The cost of prosecution and maintenance of criminals, and
the expense of the police involves an annual outlay of £4,437,000.
This, however, is small compared with the tax and toll which this
predatory horde inflicts upon the community on which it is quartered.
To the loss caused by the actual picking and stealing must be added
that of the unproductive labour of nearly 65,000 adults. Dependent
upon these criminal adults must be at least twice as many women
and children, so that it is probably an under-estimate to say that this
list of criminals and semi-criminals represents a population of at least
200,000, who all live more or less at the expense of society.

Every year, in the Metropolitan district alone, 66,100 persons are
arrested, of whom 444 are arrested for trying to commit suicide—life
having become too unbearable a burden. This immense population
is partially, no doubt, bred to prison, the same as other people are
bred to the army and to the bar. The hereditary criminal is by no
means confined to India, although it is only in that country that they
have the engaging simplicity to describe themselves frankly in
the census returns. But it is recruited constantly from the outside.
In many cases this is due to sheer starvation. Fathers of the Church
have laid down the law that a man who is in peril of death from
hunger is entitled to take bread wherever he can find it to keep body
and soul together. That proposition is not embodied in our
jurisprudence. Absolute despair drives many a man into the
ranks of the criminal class, who would never have fallen into the
category of criminal convicts if adequate provision had been made for
the rescue of those drifting to doom. When once he has fallen, circumstances seem to combine to keep him there. As wounded and sickly stags
are gored to death by their fellows, so the unfortunate who bears the
prison brand is hunted from pillar to post, until he despairs of ever
regaining his position, and oscillates between one prison and another
for the rest of his days. I gave in a preceding page an account of
how a man, after trying in vain to get work, fell before the temptation
to steal in order to escape starvation. Here is the sequel of that

man's story. After he had stolen he ran away, and thus describes his experiences :—

" To fly was easy. To get away from the scene required very little ingenuity, but the getting away from one suffering brought another. A straight look from a stranger, a quick step behind me, sent a chill through every nerve. The cravings of hunger had been satisfied, but it was the cravings of conscience that were clamorous now. It was easy to get away from the earthly consequences of sin, but from the fact—never. And yet it was the compulsion of circumstances that made me a criminal. It was neither from inward viciousness or choice, and how bitterly did I cast reproach on society for allowing such an alternative to offer itself—'to Steal or Starve,' but there was another alternative that here offered itself—either give myself up, or go on with the life of crime. I chose the former. I had travelled over 100 miles to get away from the scene of my theft, and I now find myself outside the station house at a place where I had put in my boyhood days.

" How many times when a lad, with wondering eyes, and a heart stirred with childhood's pure sympathy, I had watched the poor waifs from time to time led within its doors. It was my turn now. I entered the charge room, and with business-like precision disclosed my errand, viz. : that I wished to surrender myself for having committed a felony. My story was doubted. Question followed question, and confirmation must be waited. 'Why had I surrendered ? ' 'I was a rum 'un.' 'Cracked.' 'More fool than rogue.' 'He will be sorry when he mounts the wheel.' These and such like remarks were handed round concerning me. An hour passed by. An inspector enters, and announces the receipt of a telegram. 'It is all right. You can put him down. And turning to me, he said, 'They will send for you on Monday,' and then I passed into the inner ward, and a cell. The door closed with a harsh, grating clang, and I was left to face the most clamorous accuser of all—my own interior self.

" Monday morning, the door opened, and a complacent detective stood before me. Who can tell the feeling as the handcuffs closed round my wrists, and we started for town. As again the charge was entered, and the passing of another night in the cell ; then the morning of *the day* arrived. The gruff, harsh 'Come on' of the gaoler roused me, and the next moment I found myself in the prison van, gazing through the crevices of the floor, watching the stones flying as it were from beneath our feet. Soon the court-house was reached, and hustled into a common cell, I found myself amongst a crowd of boys and men, all bound for the 'dock.' One by one the names are called, and the crowd is gradually thinning down, when the announcement of my own name fell on my startled ear, and I found myself stumbling up the stairs, and finding myself in daylight and the 'dock.' What a terrible ordeal it was. The ceremony was brief enough; 'Have you anything to say?' 'Don't interrupt his Worship prisoner! 'Give over talking!' 'A month's hard labour.' This is about all I heard, or at any rate realised, until a vigorous push landed me into the presence of the officer who booked the sentence, and then off I went to gaol. I need not linger over the formalities of the reception. A nightmare seemed to have settled upon me as I passed into the interior of the correctional.

"I resigned my name, and I seemed to die to myself for henceforth. 332B disclosed my identity to myself and others.

"Through all the weeks that followed I was like one in a dream. Meal times, resting hours, as did every other thing, came with clock-like precision. At times I thought my mind had gone—so dull, so callous, so weary appeared the organs of the brain. The harsh orders of the gaolers; the droning of the chaplain in the chapel; the enquiries of the chief warder or the governor in their periodical visits,—all seemed so meaningless.

"As the day of my liberation drew near, the horrid conviction that circumstances would perhaps compel me to return to prison haunted me, and so helpless did I feel at the prospects that awaited me outside, that I dreaded release, which seemed but the facing of an unsympathetic world. The day arrived, and, strange as it may sound, it was with regret that I left my cell. It had become my home, and no home waited me outside.

"How utterly crushed I felt; feelings of companionship had gone out to my unfortunate fellow-prisoners, whom I had seen daily, but the sound of whose voices I had never heard, whilst outside friendships were dead, and companionships were for ever broken, and I felt as an outcast of society, with the mark of 'gaol bird' upon me, that I must cover my face, and stand aside and cry 'unclean.' Such were my feelings.

"The morning of discharge came, and I am once more on the streets. My scanty means scarcely sufficient for two days' least needs. Could I brace myself to make another honest endeavour to start afresh? Try, indeed, I did. I fell back upon my antecedents, and tried to cut the dark passage out of my life, but straight came the questions to me at each application for employment, 'What have you been doing lately?' 'Where have you been living?' If I evaded the question it caused doubt; if I answered, the only answer I could give was 'in gaol,' and that settled my chances.

"What, a comedy, after all, it appeared. I remember the last words of the chaplain before leaving the prison, cold and precise in their officialism: 'Mind you never come back here again, young man.' And now, as though in response to my earnest effort to keep from going to prison, society, by its actions, cried out, 'Go back to gaol. There are honest men enough to do our work without such as you.'

"Imagine, if you can, my condition. At the end of a few days, black despair had wrapt itself around every faculty of mind and body. Then followed several days and nights with scarcely a bit of food or a resting-place. I prowled the streets like a dog, with this difference, that the dog has the chance of helping himself, and I had not. I tried to forecast how long starvation's fingers would be in closing round the throat they already gripped. So indifferent was I alike to man or God, as I waited for the end."

In this dire extremity the writer found his way to one of our Shelters, and there found God and friends and hope, and once more got his feet on to the ladder which leads upward from the black gulf of starvation to competence and character, and usefulness and heaven.

As he was then, however, there are hundreds—nay, thousands—now. Who will give these men a helping hand? What is to be done with them? Would it not be more merciful to kill them off at once instead of sternly crushing them out of all semblance of honest manhood? Society recoils from such a short cut. Her virtuous scruples reminds me of the subterfuge by which English law evaded the veto on torture. Torture was forbidden, but the custom of placing an obstinate witness under a press and slowly crushing him within a hairbreadth of death was legalised and practised. So it is to-day. When the criminal comes out of gaol the whole world is often but a press whose punishment is sharp and cruel indeed. Nor can the victim escape even if he opens his mouth and speaks.

CHAPTER VIII.

THE CHILDREN OF THE LOST.

Whatever may be thought of the possibility of doing anything with the adults, it is universally admitted that there is hope for the children. " I regard the existing generation as lost," said a leading Liberal statesman. " Nothing can be done with men and women who have grown up under the present demoralising conditions. My only hope is that the children may have a better chance. Education will do much." But unfortunately the demoralising circumstances of the children are not being improved—are, indeed, rather, in many respects, being made worse. The deterioration of our population in large towns is one of the most undisputed facts of social economics. The country is the breeding ground of healthy citizens. But for the constant influx of Countrydom, Cockneydom would long ere this have perished. But unfortunately the country is being depopulated. The towns, London especially, are being gorged with undigested and indigestible masses of labour, and, as the result, the children suffer grievously.

The town-bred child is at a thousand disadvantages compared with his cousin in the country. But every year there are more town-bred children and fewer cousins in the country. To rear healthy children you want first a home; secondly, milk; thirdly, fresh air; and fourthly, exercise under the green trees and blue sky. All these things every country labourer's child possesses, or used to possess. For the shadow of the City life lies now upon the fields, and even in the remotest rural district the labourer who tends the cows is often denied the milk which his children need. The regular demand of the great towns forestalls the claims of the labouring hind. Tea and slops and beer take the place of milk, and the bone and sinew of the next generation are sapped from the cradle. But the country child, if he has nothing but skim milk, and only a little of that, has at least plenty of exercise in the fresh air. He has healthy human relations with his neighbours. He is looked after, and in some sort of fashion brought into contact with the life of the hall, the vicarage, and the farm. He lives a natural life amid the birds and trees and growing crops and the animals of the fields. He is not a mere human ant, crawling on the granite pavement of a great urban ants nest, with an unnaturally developed nervous system and a sickly constitution.

But, it will be said, the child of to-day has the inestimable advantage of Education. No; he has not. Educated the children are not. They are pressed through " standards," which exact a certain acquaintance with A B C and pothooks and figures, but educated they are not in the sense of the development of their latent capacities so as to make them capable for the discharge of their duties in life. The new generation can read, no doubt. Otherwise, where would be the sale of " Sixteen String Jack," " Dick Turpin," and the like? But take the girls. Who can pretend that the girls whom our schools are now turning out are half as well educated for the work of life as their grandmothers were at the same age? How many of all these mothers of the future know how to bake a loaf or wash their clothes? Except minding the baby—a task that cannot be evaded—what domestic training have they received to qualify them for being in the future the mothers of babies themselves?

And even the schooling, such as it is, at what an expense is it often imparted! The rakings of the human cesspool are brought into the school-room and mixed up with your children. Your little ones, who never heard a foul word and who are not only innocent, but ignorant, of all the horrors of vice and sin, sit for hours side by side with little ones whose parents are habitually drunk, and play with others whose ideas of merriment are gained from the familiar spectacle of the nightly debauch by which their mothers earn the family bread. It is good, no doubt, to learn the A B C, but it is not so good that in acquiring these indispensable rudiments, your children should also acquire the vocabulary of the harlot and the corner boy. I speak only of what I know, and of that which has been brought home to me as a matter of repeated complaint by my Officers, when I say that the obscenity of the talk of many of the children of some of our public schools could hardly be outdone even in Sodom and Gomorrha. Childish innocence is very beautiful ; but the bloom is soon destroyed, and it is a cruel awakening for a mother to discover that her tenderly nurtured boy, or her carefully guarded daughter, has been initiated by a companion into the mysteries of abomination that are concealed in the phrase— a house of ill-fame.

The home is largely destroyed where the mother follows the father into the factory, and where the hours of labour are so long that they have no time to see their children. The omnibus drivers of London, for instance, what time have they for discharging the daily duties of parentage to their little ones ? How can a man who is on his omnibus from fourteen to sixteen hours a day have time to be a father to his children in any sense of the word ? He has hardly a chance to see them except when they are asleep. Even the Sabbath, that blessed institution which is one of the sheet anchors of human exist- ence, is encroached upon. Many of the new industries which have been started or developed since I was a boy ignore man's need of one day's rest in seven. The railway, the post-office, the tramway all compel some of their employés to be content with less than the divinely appointed minimum of leisure. In the country darkness restores the labouring father to his little ones. In the town gas and the electric light enables the employer to rob the children of the whole of their father's waking hours, and in some cases he takes the mother's also. Under some of the conditions of modern industry, children are not so much born into a home as they are spawned into the world like fish, with the results which we see.

The decline of natural affection follows inevitably from the sub- stitution of the fish relationship for that of the human. A father who never dandles his child on his knee cannot have a very keen sense of the responsibilities of paternity. In the rush and pressure of our competitive City life, thousands of men have not time to be fathers. Sires, yes ; fathers, no. It will take a good deal of school- master to make up for that change. If this be the case, even with the children constantly employed, it can be imagined what kind of a home life is possessed by the children of the tramp, the odd jobber, the thief, and the harlot. For all these people have children, although they have no homes in which to rear them. Not a bird in all the woods or fields but prepares some kind of a nest in which to hatch and rear its young, even if it be but a hole in the sand or a few crossed sticks in the bush. But how many young ones amongst our people are hatched before any nest is ready to receive them?

Think of the multitudes of children born in our workhouses, children of whom it may be said " they are conceived in sin and shapen in iniquity," and, as a punishment of the sins of the parents, branded from birth as bastards, worse than fatherless, homeless, and friendless, "damned into an evil world," in which even those who have all the advantages of a good parentage and a careful training find it hard enough to make their way. Sometimes, it is true, the passionate love of the deserted mother for the child which has been the visible symbol and the terrible result of her undoing stands between the little one and all its enemies. But think how often the mother regards the advent of her child with loathing and horror; how the discovery that she is about to become a mother affects her like a nightmare ; and how nothing but the dread of the hangman's rope keeps her from strangling the babe on the very hour of its birth. What chances has such a child? And there are many such.

In a certain country that I will not name there exists a scientifically arranged system of infanticide cloaked under the garb of philanthropy. Gigantic foundling establishments exist in its principal cities, where every comfort and scientific improvement is provided for the deserted children, with the result that one-half of them die. The mothers are spared the crime. The State assumes the responsibility. We do something like that here, but our foundling asylums are the Street, the Workhouse, and the Grave. When an English Judge tells us, as Mr. Justice Wills did the other day, that there were any number of parents who would kill their children for a few pounds' insurance money, we can form some idea of the horrors of the existence into which many of the children of this highly favoured land are ushered at their birth.

The overcrowded homes of the poor compel the children to witness everything. Sexual morality often comes to have no meaning to them. Incest is so familiar as hardly to call for remark. The bitter poverty of the poor compels them to leave their children half fed. There are few more grotesque pictures in the history of civilisation than that of the compulsory attendance of children at school, faint with hunger because they had no breakfast, and not sure whether they would even secure a dry crust for dinner when their morning's quantum of education had been duly imparted. Children thus hungered, thus housed, and thus left to grow up as best they can without being fathered or mothered,

E

are not, educate them as you will, exactly the most promising material for the making of the future citizens and rulers of the Empire.

What, then, is the ground for hope that if we leave things alone the new generation will be better than their elders ? To me it seems that the truth is rather the other way. The lawlessness of our lads, the increased license of our girls, the general shiftlessness from the home-making point of view of the product of our factories and schools are far from reassuring. Our young people have never learned to obey. The fighting gangs of half-grown lads in Lisson Grove, and the scuttlers of Manchester are ugly symptoms of a social condition that will not grow better by being left alone.

It is the home that has been destroyed, and with the home the home-like virtues. It is the dis-homed multitude, nomadic, hungry, that is rearing an undisciplined population, cursed from birth with hereditary weakness of body and hereditary faults of character. It is idle to hope to mend matters by taking the children and bundling them up in barracks. A child brought up in an institution is too often only half-human, having never known a mother's love and a father's care. To men and women who are without homes, children must be more or less of an incumbrance. Their advent is regarded with impatience, and often it is averted by crime. The unwelcome little stranger is badly cared for, badly fed, and allowed every chance to die. Nothing is worth doing to increase his chances of living that does not Reconstitute the Home. But between us and that ideal how vast is the gulf ! It will have to be bridged, however, if anything practical is to be done.

CHAPTER IX.

IS THERE NO HELP?

It may be said by those who have followed me to this point that while it is quite true that there are many who are out of work, and not less true that there are many who sleep on the Embankment and elsewhere, the law has provided a remedy, or if not a remedy, at least a method, of dealing with these sufferers which is sufficient. Indeed, in certain circles one is not only advised that the existing state of things is inevitable, but assured that no further machinery is necessary to deal with it. All that is needed in this direction is already in working order, and to create any further machinery, or to spend any more money, will do more harm than good.

Now, what is the existing machinery by which Society, whether through the organisation of the State, or by individual endeavour, attempts to deal with the submerged residuum? I had intended at one time to have devoted considerable space to the description of the existing agencies, together with certain observations which have been forcibly impressed upon my mind as to their failure and its cause. The necessity, however, of subordinating everything to the supreme purpose of this book, which is to endeavour to show how light can be let into the heart of Darkest England, compels me to pass rapidly over this department of the subject, merely glancing as I go at the well-meaning, but more or less abortive, attempts to cope with this great and appalling evil.

The first place must naturally be given to the administration of the Poor Law. Legally the State accepts the responsibility of providing food and shelter for every man, woman, or child who is utterly destitute. This responsibility it, however, practically shirks by the imposition of conditions on the claimants of relief that are hateful and repulsive, if not impossible. As to the method of Poor

Law administration in dealing with inmates of workhouses or in the distribution of outdoor relief, I say nothing. Both of these raise great questions which lie outside my immediate purpose. All that I need to do is to indicate the limitations—it may be the necessary limitations—under which the Poor Law operates. No Englishman can come upon the rates so long as he has anything whatever left to call his own. When long-continued destitution has been carried on to the bitter end, when piece by piece every article of domestic furniture has been sold or pawned, when all efforts to procure employment have failed, and when you have nothing left except the clothes in which you stand, then you can present yourself before the relieving officer and secure your lodging in the workhouse, the administration of which varies infinitely according to the disposition of the Board of Guardians under whose control it happens to be.

If, however, you have not sunk to such despair as to be willing to barter your liberty for the sake of food, clothing, and shelter in the Workhouse, but are only temporarily out of employment, seeking work, then you go to the Casual Ward. There you are taken in, and provided for on the principle of making it as disagreeable as possible for yourself, in order to deter you from again accepting the hospitality of the rates,—and of course in defence of this a good deal can be said by the Political Economist. But what seems utterly indefensible is the careful precautions which are taken to render it impossible for the unemployed Casual to resume promptly after his night's rest the search for work. Under the existing regulations, if you are compelled to seek refuge on Monday night in the Casual Ward, you are bound to remain there at least till Wednesday morning.

The theory of the system is this, that individuals casually poor and out of work, being destitute and without shelter, may upon application receive shelter for the night, supper and a breakfast, and in return for this, shall perform a task of work, not necessarily in repayment for the relief received, but simply as a test of their willingness to work for their living. The work given is the same as that given to felons in gaol, oakum-picking and stone-breaking.

The work, too, is excessive in proportion to what is received. Four pounds of oakum is a great task to an expert and an old hand. To a novice it can only be accomplished with the greatest difficulty, if indeed it can be done at all. It is even in excess of the amount demanded from a criminal in gaol.

The stone-breaking test is monstrous. Half a ton of stone from any man in return for partially supplying the cravings of hunger is an outrage which, if we read of as having occurred in Russia or Siberia, would find Exeter Hall crowded with an indignant audience, and Hyde Park filled with strong oratory. But because this system exists at our own doors, very little notice is taken of it. These tasks are expected from all comers, starved, ill-clad, half-fed creatures from the streets, foot-sore and worn out, and yet unless it is done, the alternative is the magistrate and the gaol. The old system was bad enough, which demanded the picking of one pound of oakum. As soon as this task was accomplished, which generally kept them till the middle of next day, it was thus rendered impossible for them to seek work, and they were forced to spend another night in the ward. The Local Government Board, however, stepped in, and the Casual was ordered to be detained for the whole day and the second night, the amount of labour required from him being increased four-fold.

Under the present system, therefore, the penalty for seeking shelter from the streets is a whole day and two nights, with an almost impossible task, which, failing to do, the victim is liable to be dragged before a magistrate and committed to gaol as a rogue and vagabond, while in the Casual Ward their treatment is practically that of a criminal. They sleep in a cell with an apartment at the back, in which the work is done, receiving at night half a pound of gruel and eight ounces of bread, and next morning the same for breakfast, with half a pound of oakum and stones to occupy himself for a day.

The beds are mostly of the plank type, the coverings scant, the comfort *nil.* Be it remembered that this is the treatment meted out to those who are supposed to be Casual poor, in temporary difficulty, walking from place to place seeking some employment.

The treatment of the women is as follows : Each Casual has to stay in the Casual Wards two nights and one day, during which time they have to pick 2 lb. of oakum or go to the wash-tub and work out the time there. While at the wash-tub they are allowed to wash their own clothes, but not otherwise. If seen more than once in the same Casual Ward, they are detained three days by order of the inspector each time seen, or if sleeping twice in the same month the master of the ward has power to detain them three days. There are four inspectors who visit different Casual Wards ; and if the Casual is seen by any of the inspectors (who in turn visit all the Casual Wards) at any of the wards they have previously visited they are detained three days in each one. The inspector, who is a male person, visits the wards at all unexpected hours, even visiting while the females are in bed. The beds are in some wards composed of straw and two rugs, in others cocoanut fibre and two rugs. The Casuals rise at 5.45 a.m. and go to bed 7 p.m. If they do not finish picking their oakum before 7 p.m., they stay up till they do. If a Casual does not come to the ward before 12.30, midnight, they keep them one day extra. The way in which this operates, however, can be best understood by the following statements, made by those who have been in Casual Wards, and who can, therefore, speak from experience as to how the system affects the individual :—

J. C. knows Casual Wards pretty well. Has been in St. Giles, Whitechapel, St. George's, Paddington, Marylebone, Mile End. They vary a little in detail, but as a rule the doors open at 6; you walk in; they tell you what the work is, and that if you fail to do it, you will be liable to imprisonment. Then you bathe. Some places the water is dirty. Three persons as a rule wash in one water. At Whitechapel (been there three times) it has always been dirty; also at St. George's. I had no bath at Mile End; they were short of water. If you complain they take no notice. You then tie your clothes in a bundle, and they give you a nightshirt. At most places they serve supper to the men, who have to go to bed and eat it there. Some beds are in cells; some in large rooms. You get up at 6 a.m. and do the task. The amount of stonebreaking is too much; and the oakum-picking is also heavy. The food differs. At St. Giles, the gruel left over-night is boiled up for breakfast, and is consequently sour; the bread is puffy, full of holes, and don't weigh the regulation amount. Dinner is only 8 ounces of bread and $1\frac{1}{2}$ ounce of cheese, and if that's short, how can anybody do their work? They will give you water to drink if you ring the cell bell for it, that is, they will tell you to wait, and bring it in about half an hour. There are a good lot of "moochers" go to Casual Wards, but there are large numbers of men who only want work.

J. D. ; age 25; Londoner; can't get work, tried hard; been refused work several times on account of having no settled residence; looks suspicious, they think, to have "no home." Seems a decent, willing man. Had two pennyworth of soup this morning, which has lasted all day. Earned 1s. 6d. yesterday, bill distributing, nothing the day before. Been in good many London Casual Wards. Thinks they are no good, because they keep him all day, when he might be seeking work. Don't want shelter in day time, wants work. If he goes in twice in a month to the same Casual Ward, they detain him four days. Considers the food decidedly insufficient to do the required amount of work. If the work is not done to time, you are liable to 21 days' imprisonment. Get badly treated some places, especially where there is a bullying superintendent. Has done 21 days for absolutely refusing to do the work on such low diet, when unfit. Can't get justice, doctor always sides with superintendent.

J. S.; odd jobber. Is working at board carrying, when he can get it. There's quite a rush for it at 1s. 2d. a day. Carried a couple of parcels yesterday, got 5d. for them; also had a bit of bread and meat given him by a working man, so altogether had an excellent day. Sometimes goes all day without food, and plenty more do the same. Sleeps on Embankment, and now and then in Casual Ward. Latter is clean and comfortable enough, but they keep you in all day; that means no chance of getting work. Was a clerk once, but got out of a job, and couldn't get another; there are so many clerks.

"A Tramp" says: "I've been in most Casual Wards in London; was in the one in Macklin Street, Drury Lane, last week. They keep you two nights and a day, and more than that if they recognise you. You have to break 10 cwt. of stone, or pick four pounds of oakum. Both are hard. About thirty a night go to Macklin Street. The food is 1 pint gruel and 6 oz. bread for breakfast; 8 oz. bread and 1½ oz. cheese for dinner; tea same as breakfast. No supper. It is not enough to do the work on. Then you are obliged to bathe, of course; sometimes three will bathe in one water, and if you complain they turn nasty, and ask if you are come to a palace. Mitcham Workhouse I've been in; grub is good; 1½ pint gruel and 8 oz. bread for breakfast, and same for supper.

F. K. W.; baker. Been board-carrying to-day, earned one shilling, hours 9 till 5. I've been on this kind of life six years. Used to work in a bakery, but had congestion of the brain, and couldn't stand the heat. I've been in about every Casual Ward in England. They treat men too harshly. Have to work very hard, too. Has had to work whilst really unfit. At Peckham (known as Camberwell) Union, was quite unable to do it through weakness, and appealed to the doctor, who, taking the part of the other officials, as usual, refused to allow him to forego the work. Cheeked the doctor, telling him he didn't understand his work; result, got three days' imprisonment. Before going to a Casual Ward at all, I spent seven consecutive nights on the Embankment, and at last went to the Ward.

The result of the deliberate policy of making the night refuge for the unemployed labourer as disagreeable as possible, and of placing as many obstacles as possible in the way of his finding work the following day, is, no doubt, to minimise the number of Casuals, and without question succeeds. In the whole of London the number of Casuals in the wards at night is only 1,136. That is to say, the conditions which are imposed are so severe, that the majority of the Out-of-Works prefer to sleep in the open air, taking their chance of the inclemency and mutability of our English weather, rather than go through the experience of the Casual Ward.

It seems to me that such a mode of coping with distress does not so much meet the difficulty as evade it. It is obvious that an apparatus, which only provides for 1,136 persons per night, is utterly unable to deal with the numbers of the homeless Out-of-Works. But if by some miracle we could use the Casual Wards as a means of providing for all those who are seeking work from day to day, without a place in which to lay their heads, save the kerbstone of the pavement or the back of a seat on the Embankment, they would utterly fail to have any appreciable effect upon the mass of human misery with which we have to deal. For this reason; the administration of the Casual Wards is mechanical, perfunctory, and formal. Each of the Casuals is to the Officer in Charge merely one Casual the more. There is no attempt whatever to do more than provide for them merely the indispensable requisites of existence. There has never been any attempt to treat them as human beings, to deal with them as individuals, to appeal to their hearts, to help them on their legs again. They are simply units, no more thought of and cared for than if they were so many coffee beans passing through a coffee mill; and as the net result of all my experience and observation of men and things, I must assert unhesitatingly that anything which dehumanises the individual, anything which treats a man as if he were only a number of a series or a cog in a wheel, without any regard to the character, the aspirations, the temptations, and the idiosyncrasies of the man, must utterly fail as a remedial agency. The Casual Ward, at the best, is merely a squalid resting place for the Casual in his downward career. If anything is to be done for these men, it must be done by other agents than those which prevail in the administration of the Poor Laws.

The second method in which Society endeavours to do its duty to the lapsed masses is by the miscellaneous and heterogeneous efforts which are clubbed together under the generic head of Charity. Far be it from me to say one word in disparagement of any effort that is prompted by a sincere desire to alleviate the misery of our fellow creatures, but the most charitable are those who most deplore the utter failure which has, up till now, attended all their efforts to do more than temporarily alleviate pain, or effect an occasional improvement in the condition of individuals.

There are many institutions, very excellent in their way, without which it is difficult to see how society could get on at all, but when they have done their best there still remains this great and appalling mass of human misery on our hands, a perfect quagmire of Human Sludge. They may ladle out individuals here and there, but to drain the whole bog is an effort which seems to be beyond the imagination of most of those who spend their lives in philanthropic work. It is no doubt better than nothing to take the individual and feed him from day to day, to bandage up his wounds and heal his diseases ; but you may go on doing that for ever, if you do not do more than that ; and the worst of it is that all authorities agree that if you only do that you will probably increase the evil with which you are attempting to deal, and that you had much better let the whole thing alone.

There is at present no attempt at Concerted Action. Each one deals with the case immediately before him, and the result is what might be expected ; there is a great expenditure, but the gains are, alas ! very small. The fact, however, that so much is subscribed for the temporary relief and the mere alleviation of distress justifies my confidence that if a Practical Scheme of dealing with this misery in a permanent, comprehensive fashion be discovered, there will be no lack of the sinews of war. It is well, no doubt, sometimes to administer an anæsthetic, but the Cure of the Patient is worth ever so much more, and the latter is the object which we must constantly set before us in approaching this problem.

The third method by which Society professes to attempt the re-clamation of the lost is by the rough, rude surgery of the Gaol. Upon this a whole treatise might be written, but when it was finished it would be nothing more than a demonstration that our Prison system has practically missed aiming at that which should be the first essential of every system of punishment. It is not Refor-matory, it is not worked as if it were intended to be Reformatory. It is punitive, and only punitive. The whole administration needs to be reformed from top to bottom in accordance with this fundamental prin-ciple, viz., that while every prisoner should be subjected to that measure of punishment which shall mark a due sense of his crime both to himself and society, the main object should be to rouse in his mind the desire to lead an honest life ; and to effect that change in his disposition and character which will send him forth to put that desire into practice. At present, every Prison is more or less a Training School for Crime, an introduction to the society of criminals, the petrifaction of any lingering human feeling and a very Bastille of Despair. The prison brand is stamped upon those who go in, and that so deeply, that it seems as if it clung to them for life. To enter Prison once, means in many cases an almost certain return there at an early date. All this has to be changed, and will be, when once the work of Prison Reform is taken in hand by men who understand the subject, who believe in the reformation of human nature in every form which its depravity can assume, and who are in full sympathy with the class for whose benefit they labour; and when those charged directly with the care of criminals seek to work out their regeneration in the same spirit.

The question of Prison Reform is all the more important because it is only by the agency of the Gaol that Society attempts to deal with its hopeless cases. If a woman, driven mad with shame, flings herself into the river, and is fished out alive, we clap her into Prison on a charge of attempted suicide. If a man, despairing of work and gaunt with hunger, helps himself to food, it is to the same reformatory agency that he is forthwith subjected. The rough and ready surgery with which we deal with our social patients recalls the simple method of the early physicians. The tradition still lingers among old people of doctors who prescribed bleeding for every ailment, and of keepers of asylums whose one idea of ministering to a mind diseased was to put the body into a strait waistcoat. Modern science laughs to scorn these simple "remedies" of an unscientific age, and declares that they were, in most cases, the most efficacious means of aggravating the disease they professed to cure. But in social maladies we are still in the age of the blood-letter and the strait waistcoat. The Gaol is our specific for Despair. When all else fails Society will always undertake to feed, clothe, warm, and house a man, if only he will commit a crime. It will do it also in such a fashion as to render it no temporary help, but a permanent necessity.

Society says to the individual: "To qualify for free board and lodging you must commit a crime. But if you do you must pay the price. You must allow me to ruin your character, and doom you for the rest of your life to destitution, modified by the occasional successes of criminality. You shall become the Child of the State, on condition that we doom you to a temporal perdition, out of which you will never be permitted to escape, and in which you will always be a charge upon our resources and a constant source of anxiety and inconvenience to the authorities. I will feed you, certainly, but in return you must permit me to damn you." That surely ought not to be the last word of Civilised Society.

"Certainly not," say others. "Emigration is the true specific. The waste lands of the world are crying aloud for the application of surplus labour. Emigration is the panacea." Now I have no objection to emigration. Only a criminal lunatic could seriously object to the transference of hungry Jack from an overcrowded shanty— where he cannot even obtain enough bad potatoes to dull the ache behind his waistcoat, and is tempted to let his child die for the sake of the insurance money—to a land flowing with milk and honey, where he can eat meat three times a day and where a man's children are his wealth. But you might as well lay a new-born child naked in the middle of a new-sown field in March, and expect it to live and thrive, as expect emigration to produce successful results on the lines which some lay down. The child, no doubt, has within it latent capacities which, when years and training have done their work, will enable him to reap a harvest from a fertile soil, and the new sown field will be covered with golden grain in August. But these facts will not enable the infant to still its hunger with the clods of the earth in the cold spring time. It is just like that with emigration. It is simply criminal to take a multitude of untrained men and women and land them penniless and helpless on the fringe of some new continent. The result of such proceedings we see in the American cities; in the degradation of their slums, and in the hopeless demoralisation of thousands who, in their own country, were living decent, industrious lives.

A few months since, in Paramatta, in New South Wales, a young man who had emigrated with a vague hope of mending his fortunes, found himself homeless, friendless, and penniless. He was a clerk. They wanted no more clerks in Paramatta. Trade was dull, employment was scarce, even for trained hands. He went about from day to day seeking work and finding none. At last he came to the end of all his resources. He went all day without food; at night he slept as best he could. Morning came, and he was hopeless. All next day passed without a meal. Night came. He could not sleep. He wandered about restlessly. At last, about midnight, an idea seized him. Grasping a brick, he deliberately walked up to a

jeweller's window, and smashed a hole through the glass. He
made no attempt to steal anything. He merely smashed the
pane and then sat down on the pavement beneath the window,
waiting for the arrival of the policeman. He waited some hours ;
but at last the constable arrived. He gave himself up, and was
marched off to the lock-up. " I shall at least have something to eat
now," was the reflection. He was right. He was sentenced to
one year's imprisonment, and he is in gaol at this hour. This very
morning he received his rations, and at this very moment he is
lodged, and clothed and cared for at the cost of the rates and taxes.
He has become the child of the State, and, therefore, one of the
socially damned. Thus emigration itself, instead of being an
invariable specific, sometimes brings us back again to the gaol door.

Emigration, by all means. But whom are you to emigrate ?
These girls who do not know how to bake ? These lads who never
handled a spade ? And where are you to emigrate them ? Are
you going to make the Colonies the dumping ground of your human
refuse ? On that the colonists will have something decisive to say,
where there are colonists ; and where there are not, how are you
to feed, clothe, and employ your emigrants in the uninhabited
wilderness ? Immigration, no doubt, is the making of a colony,
just as bread is the staff of life. But if you were to cram a stomach
with wheat by a force-pump you would bring on such a fit of
indigestion that unless your victim threw up the indigestible mass
of unground, uncooked, unmasticated grain he would never want
another meal. So it is with the new colonies and the surplus labour
of other countries.

Emigration is in itself not a panacea. Is Education? In one
sense it may be, for Education, the developing in a man of all his
latent capacities for improvement, may cure anything and everything.
But the Education of which men speak when they use the term, is
mere schooling. No one but a fool would say a word against school
teaching. By all means let us have our children educated. But
when we have passed them through the Board School Mill we have
enough experience to see that they do not emerge the renovated
and regenerated beings whose advent was expected by those who
passed the Education Act. The " scuttlers " who knife inoffensive
persons in Lancashire, the fighting gangs of the West of London,
belong to the generation that has enjoyed the advantage of Compulsory
Education. Education, book-learning and schooling will not

solve the difficulty. It helps, no doubt. But in some ways it aggravates it. The common school to which the children of thieves and harlots and drunkards are driven, to sit side by side with our little ones, is often by no means a temple of all the virtues. It is sometimes a university of all the vices. The bad infect the good, and your boy and girl come back reeking with the contamination of bad associates, and familiar with the coarsest obscenity of the slum. Another great evil is the extent to which our Education tends to overstock the labour market with material for quill-drivers and shopmen, and gives our youth a distaste for sturdy labour. Many of the most hopeless cases in our Shelters are men of considerable education. Our schools help to enable a starving man to tell his story in more grammatical language than that which his father could have employed, but they do not feed him, or teach him where to go to get fed. So far from doing this they increase the tendency to drift into those channels where food is least secure, because employment is most uncertain, and the market most overstocked.

"Try Trades Unionism," say some, and their advice is being widely followed. There are many and great advantages in Trades Unionism. The fable of the bundle of sticks is good for all time. The more the working people can be banded together in voluntary organisations, created and administered by themselves for the protection of their own interests, the better—at any rate for this world—and not only for their own interests, but for those of every other section of the community. But can we rely upon this agency as a means of solving the problems which confront us? Trades Unionism has had the field to itself for a generation. It is twenty years since it was set free from all the legal disabilities under which it laboured. But it has not covered the land. It has not organised all skilled labour. Unskilled labour is almost untouched. At the Congress at Liverpool only one and a half million workmen were represented. Women are almost entirely outside the pale. Trade Unions not only represent a fraction of the labouring classes, but they are, by their constitution, unable to deal with those who do not belong to their body. What ground can there be, then, for hoping that Trades Unionism will by itself solve the difficulty? The most experienced Trades Unionists will be the first to admit that any scheme which could deal adequately with the out-of-works and others who hang on to their skirts and form the recruiting ground of blacklegs and embarrass them in every way, would be, of all others that which would be most beneficial to Trades Unionism. The same may be said about Co-operation. Personally, I am a strong believer in Co-operation, but it must be Co-operation based on the spirit of benevolence. I don't see how any pacific re-adjustment of the social and economic relations between classes in this country can be effected except by the gradual substitution of co-operative associations for the present wages system. As you will see in subsequent chapters, so far from there being anything in my proposals that would militate in any way against the ultimate adoption of the co-operative solution of the question, I look to Co-operation as one of the chief elements of hope in the future. But we have not to deal with the ultimate future, but with the immediate present, and for the evils with which we are dealing the existing co-operative organisations do not and cannot give us much help.

Another—I do not like to call it specific ; it is only a name, a mere mockery of a specific—so let me call it another suggestion made when discussing this evil, is Thrift. Thrift is a great virtue no doubt. But how is Thrift to benefit those who have nothing ? What is the use of the gospel of Thrift to a man who had nothing to eat yesterday, and has not threepence to-day to pay for his lodging to-night ? To live on nothing a day is difficult enough, but to save on it would beat the cleverest political economist that ever lived. I admit without hesitation that any Scheme which weakened the incentive to Thrift would do harm. But it is a mistake to imagine that social damnation is an incentive to Thrift. It operates least where its force ought to be most felt. There is no fear that any Scheme that we can devise will appreciably diminish the deterrent influences which dispose a man to save. But it is idle wasting time upon a plea that is only brought forward as an excuse for inaction. Thrift is a great virtue, the inculcation of which must be constantly kept in view by all those who are attempting to educate and save the people. It is not in any sense a specific for the salvation of the lapsed and the lost. Even among the most wretched of the very poor, a man must have an object and a hope before he will save a halfpenny. " Let us eat and drink, for to-morrow we perish," sums up the philosophy of those who have no hope. In the thriftiness of the French peasant we see that the temptation of eating and drinking is capable of being resolutely subordinated to the superior claims of the accumulation of a dowry for the daughter, or for the acquisition of a little more land for the son.

Of the schemes of those who propose to bring in a new heaven and a new earth by a more scientific distribution of the pieces of gold and silver in the trouser pockets of mankind, I need not say anything here. They may be good or they may not. I say nothing against any short cut to the Millennium that is compatible with the Ten Commandments. I intensely sympathise with the aspirations that lie behind all these Socialist dreams. But whether it is Henry George's Single Tax on Land Values, or Edward Bellamy's Nationalism, or the more elaborate schemes of the Collectivists, my attitude towards them all is the same. What these good people want to do, I also want to do. But I am a practical man, dealing with the actualities of to-day. I have no preconceived theories, and I flatter myself I am singularly free from prejudices. I am ready to sit at the feet of any who will show me any good. I keep my mind open on all these subjects ; and am quite prepared to hail with open arms any Utopia that is offered me. But it must be within range of my finger-tips. It is of no use to me if it is in the clouds. Cheques on the Bank of Futurity I accept gladly enough as a free gift, but I can hardly be expected to take them as if they were current coin, or to try to cash them at the Bank of England.

It may be that nothing will be put permanently right until everything has been turned upside down. There are certainly so many things that need transforming, beginning with the heart of each individual man and woman, that I do not quarrel with any Visionary when in his intense longing for the amelioration of the condition of mankind he lays down his theories as to the necessity for radical change, however impracticable they may appear to me. But this is the question. Here at our Shelters last night were a thousand hungry, workless people. I want to know what to do with them ? Here is John Jones, a stout stalwart labourer in rags, who has not had one square meal for a month, who has been hunting for work that will enable him to keep body and soul together, and hunting in vain. There he is in his hungry raggedness, asking for work that he may live, and not die of sheer starvation in the midst of the wealthiest city in the world. What is to be done with John Jones ?

The individualist tells me that the free play of the Natural Laws governing the struggle for existence will result in the Survival of the Fittest, and that in the course of a few ages, more or less, a much nobler type will be evolved. But meanwhile what is to become of John.

Jones? The Socialist tells me that the great Social Revolution is looming large on the horizon. In the good time coming, when wealth will be re-distributed and private property abolished, all stomachs will be filled and there will be no more John Jones' impatiently clamouring for opportunity to work that they may not die. It may be so, but in the meantime here is John Jones growing more impatient than ever because hungrier, who wonders if he is to wait for a dinner until the Social Revolution has arrived. What are we to do with John Jones? That is the question. And to the solution of that question none of the Utopians give me much help. For practical purposes these dreamers fall under the condemnation they lavish so freely upon the conventional religious people who relieve themselves of all anxiety for the welfare of the poor by saying that in the next world all will be put right. This religious cant, which rids itself of all the importunity of suffering humanity by drawing unnegotiable bills payable on the other side of the grave, is not more impracticable than the Socialistic clap-trap which postpones all redress of human suffering until after the general overturn. Both take refuge in the Future to escape a solution of the problems of the Present, and it matters little to the sufferers whether the Future is on this side of the grave or the other. Both are, for them, equally out of reach.

When the sky falls we shall catch larks. No doubt. But in the meantime?

It is the meantime—that is the only time in which we have to work It is in the meantime that the people must be fed, that their life's work must be done or left undone for ever. Nothing that I have to propose in this book, or that I propose to do by my Scheme, will in the least prevent the coming of any of the Utopias. I leave the limitless infinite of the Future to the Utopians. They may build there as they please. As for me, it is indispensable that whatever I do is founded on existing fact, and provides a present help for the actual need.

There is only one class of men who have cause to oppose the proposals which I am about to set forth. That is those, if such there be, who are determined to bring about by any and every means a bloody and violent overturn of all existing institutions. They will oppose the Scheme, and they will act logically in so doing. For the only hope of those who are the artificers of Revolution is the mass of seething discontent and misery that lies in the heart of the social system. Honestly believing that things must get worse before they get better, they build all their hopes upon the general overturn, and they resent as an indefinite postponement of the realisation of their dreams any attempt at a reduction of human misery.

The Army of the Revolution is recruited by the Soldiers of Despair. Therefore, down with any Scheme which gives men Hope. In so far as it succeeds it curtails our recruiting ground and reinforces the ranks of our Enemies. Such opposition is to be counted upon, and to be utilised as the best of all tributes to the value of our work. Those who thus count upon violence and bloodshed are too few to hinder, and their opposition will merely add to the momentum with which I hope and believe this Scheme will ultimately be enabled to surmount all dissent, and achieve, with the blessing of God, that measure of success with which I verily believe it to be charged.

PART II.—DELIVERANCE.

CHAPTER I.

A STUPENDOUS UNDERTAKING.

Such, then, is a brief and hurried survey of Darkest England, and those who have been in the depths of the enchanted forest in which wander the tribes of the despairing Lost will be the first to admit that I have in no way exaggerated its horrors, while most will assert that I have under-estimated the number of its denizens. I have, indeed, very scrupulously striven to keep my estimates of the extent of the evil within the lines of sobriety. Nothing in such an enterprise as that on which I am entering could worse befall me than to come under the reproach of sensationalism or exaggeration. Most of the evidence upon which I have relied is taken direct from the official statistics supplied by the Government Returns; and as to the rest, I can only say that if my figures are compared with those of any other writer upon this subject, it will be found that my estimates are the lowest. I am not prepared to defend the exact accuracy of my calculations, excepting so far as they constitute the minimum. To those who believe that the numbers of the wretched are far in excess of my figures, I have nothing to say, excepting this, that if the evil is so much greater than I have described, then let your efforts be proportioned to your estimate, not to mine. The great point with each of us is, not how many of the wretched exist to-day, but how few shall there exist in the years that are to come.

The dark and dismal jungle of pauperism, vice, and despair is the inheritance to which we have succeeded from the generations and centuries past, during which wars, insurrections, and internal troubles left our forefathers small leisure to attend to the well-being of the sunken tenth. Now that we have happened upon more fortunate times, let us recognise that we are our brother's keepers, and set to work, regardless of party distinctions and religious differences, to make this world of ours a little bit more like home for those whom we call our brethren.

The problem, it must be admitted, is by no means a simple one; nor can anyone accuse me in the foregoing pages of having minimised the difficulties which heredity, habit, and surroundings place in the way of its solution, but unless we are prepared to fold our arms in selfish ease and say that nothing can be done, and thereby doom those lost millions to remediless perdition in this world, to say nothing of the next, the problem must be solved in some way. But in what way? That is the question. It may tend, perhaps, to the crystallisation of opinion on this subject if I lay down, with such precision as I can command, what must be the essential elements of any scheme likely to command success.

Section i.—THE ESSENTIALS TO SUCCESS.

The first essential that must be borne in mind as governing every Scheme that may be put forward is that it must change the man when it is his character and conduct which constitute the reasons for his failure in the battle of life. No change in circumstances, no revolution in social conditions, can possibly transform the nature of man. Some of the worst men and women in the world, whose names are chronicled by history with a shudder of horror, were those who had all the advantages that wealth, education and station could confer or ambition could attain.

The supreme test of any scheme for benefiting humanity lies in the answer to the question, What does it make of the individual? Does it quicken his conscience, does it soften his heart, does it enlighten his mind, does it, in short, make more of a true man of him, because only by such influences can he be enabled to lead a human life? Among the denizens of Darkest England there are many who have found their way thither by defects of character which would under the most favourable circumstances relegate them to the same position. Hence, unless you can change their character your labour will be lost. You may clothe the drunkard, fill his purse with gold, establish him in a well-furnished home, and in three, or six, or twelve months he will once more be on the Embankment, haunted by delirium tremens, dirty, squalid, and ragged. Hence, in all cases where a man's own character and defects constitute the reasons for his fall, that character must be changed and that conduct altered if any permanent beneficial results are to be attained. If he is a drunkard, he must be made sober; if idle, he must be made industrious; if criminal, he must be made honest; if impure, he must be made clean; and if he be so deep down in vice, and has been there so long that he has lost all heart, and hope, and power to help himself, and absolutely refuses to move, he must be inspired with hope and have created within him the ambition to rise; otherwise he will never get out of the horrible pit.

Secondly : *The remedy, to be effectual, must change the circumstances of the individual when they are the cause of his wretched condition, and lie beyond his control.* Among those who have arrived at their present evil plight through faults of self-indulgence or some defect in their moral character, how many are there who would have been very differently placed to-day had their surroundings been otherwise ? Charles Kingsley puts this very abruptly where he makes the Poacher's widow say, when addressing the Bad Squire, who drew back

> " Our daughters, with base-born babies,
>   Have wandered away in their shame.
> If your misses had slept, Squire, where they did,
>   Your misses might do the same.'

Placed in the same or similar circumstances, how many of us would have turned out better than this poor, lapsed, sunken multitude ?

Many of this crowd have never had a chance of doing better ; they have been born in a poisoned atmosphere, educated in circumstances which have rendered modesty an impossibility, and have been thrown into life in conditions which make vice a second nature. Hence, to provide an effective remedy for the evils which we are deploring these circumstances must be altered, and unless my Scheme effects such a change, it will be of no use. There are multitudes, myriads, of men and women, who are floundering in the horrible quagmire beneath the burden of a load too heavy for them to bear ; every plunge they take forward lands them deeper ; some have ceased even to struggle, and lie prone in the filthy bog, slowly suffocating, with their manhood and womanhood all but perished. It is no use standing on the firm bank of the quaking morass and anathematising these poor wretches ; if you are to do them any good, you must give them another chance to get on their feet, you must give them firm foothold upon which they can once more stand upright, and you must build stepping-stones across the bog to enable them safely to reach the other side. Favourable circumstances will not change a man's heart or transform his nature, but unpropitious circumstances may render it absolutely impossible for him to escape, no matter how he may desire to extricate himself. The first step with these helpless, sunken creatures is to create the desire to escape, and then provide the means for doing so. In other words, give the man another chance.

Thirdly : *Any remedy worthy of consideration must be on a scale commensurate with the evil with which it proposes to deal.*   It is no use trying to bail out the ocean with a pint pot.   This evil is one whose victims are counted by the million.   The army of the Lost in our midst exceeds the numbers of that multitudinous host which Xerxes led from Asia to attempt the conquest of Greece.   Pass in parade those who make up the submerged tenth, count the paupers indoor and outdoor, the homeless, the starving, the criminals, the lunatics, the drunkards, and the harlots—and yet do not give way to despair !   Even to attempt to save a tithe of this host requires that we should put much more force and fire into our work than has hitherto been exhibited by anyone.   There must be no more philanthropic tinkering, as if this vast sea of human misery were contained in the limits of a garden pond.

Fourthly : *Not only must the Scheme be large enough, but it must be permanent.*   That is to say, it must not be merely a spasmodic effort coping with the misery of to-day ; it must be established on a durable footing, so as to go on dealing with the misery of to-morrow and the day after, so long as there is misery left in the world with which to grapple.

Fifthly : *But while it must be permanent*, it must also be *immediately practicable.*   Any Scheme, to be of use, must be capable of being brought into instant operation with beneficial results.

Sixthly : *The indirect features of the Scheme must not be such as to produce injury to the persons whom we seek to benefit.*   Mere charity, for instance, while relieving the pinch of hunger, demoralises the recipient; and whatever the remedy is that we employ, it must be of such a nature as to do good without doing evil at the same time.   It is no use conferring sixpennyworth of benefit on a man if, at the same time, we do him a shilling'sworth of harm.

Seventhly : *While assisting one class of the community, it must not seriously interfere with the interests of another.*   In raising one section of the fallen, we must not thereby endanger the safety of those who with difficulty are keeping on their feet.

These are the conditions by which I ask you to test the Scheme I am about to unfold.   They are formidable enough, possibly, to deter many from even attempting to do anything.   They are not of my making.   They are obvious to anyone who looks into the matter. They are the laws which govern the work of the philanthropic reformer, just as the laws of gravitation, of wind and of weather, govern the operations of the engineer. It is no use saying we could build a bridge across the Tay if the wind did not blow, or that we could build a railway across a bog if the quagmire would afford us a solid foundation. The engineer has to take into account the difficulties, and make them his starting point. The wind will blow, therefore the bridge must be made strong enough to resist it. Chat Moss will shake ; therefore we must construct a foundation in the very bowels of the bog on which to build our railway. So it is with the social difficulties which confront us. If we act in harmony with these laws we shall triumph ; but if we ignore them they will overwhelm us with destruction and cover us with disgrace.

But, difficult as the task may be, it is not one which we can neglect. When Napoleon was compelled to retreat under circumstances which rendered it impossible for him to carry off his sick and wounded, he ordered his doctors to poison every man in the hospital. A general has before now massacred his prisoners rather than allow them to escape. These Lost ones are the Prisoners of Society ; they are the Sick and Wounded in our Hospitals. What a shriek would arise from the civilised world if it were proposed to administer to-night to every one of these millions such a dose of morphine that they would sleep to wake no more. But so far as they are concerned, would it not be much less cruel thus to end their life than to allow them to drag on day after day, year after year, in misery, anguish, and despair, driven into vice and hunted into crime, until at last disease harries them into the grave ?

I am under no delusion as to the possibility of inaugurating a millennium by my Scheme; but the triumphs of science deal so much with the utilisation of waste material, that I do not despair of something effectual being accomplished in the utilisation of this waste human product. The refuse which was a drug and a curse to our manufacturers, when treated under the hands of the chemist, has been the means of supplying us with dyes rivalling in loveliness and variety the hues of the rainbow. If the alchemy of science can extract beautiful colours from coal tar, cannot Divine alchemy enable us to evolve gladness and brightness out of the agonised hearts and dark, dreary, loveless lives of these doomed myriads ? Is it too much to hope that in God's world God's children may be able to do something, if they set to work with a will, to carry out a plan of campaign against these great evils which are the nightmare of our existence ?

The remedy, it may be, is simpler than some imagine. The key to the enigma may lie closer to our hands than we have any idea of. Many devices have been tried, and many have failed, no doubt ; it is only stubborn, reckless perseverance that can hope to succeed ; it is well that we recognise this. How many ages did men try to make gunpowder and never succeeded ? They would put saltpetre to charcoal, or charcoal to sulphur, or saltpetre to sulphur, and so were ever unable to make the compound explode. But it has only been discovered within the last few hundred years that all three were needed. Before that gunpowder was a mere imagination, a phantasy of the alchemists. How easy it is to make gunpowder, now the secret of its manufacture is known !

But take a simpler illustration, one which lies even within the memory of some that read these pages. From the beginning of the world down to the beginning of this century, mankind had not found out, with all its striving after cheap and easy transport, the miraculous difference that would be brought about by laying down two parallel lines of metal. All the great men and the wise men of the past lived and died oblivious of that fact. The greatest mechanicians and engineers of antiquity, the men who bridged all the rivers of Europe, the architects who built the cathedrals which are still the wonder of the world, failed to discern what seems to us so obviously simple a proposition, that two parallel lines of rail would diminish the cost and difficulty of transport to a minimum. Without that discovery the steam engine, which has itself been an invention of quite recent years, would have failed to transform civilisation.

What we have to do in the philanthropic sphere is to find something analogous to the engineers' parallel bars. This discovery I think I have made, and hence have I written this book.

Section 2.—MY SCHEME.

What, then, is my Scheme? It is a very simple one, although in its ramifications and extensions it embraces the whole world. In this book I profess to do no more than to merely outline, as plainly and as simply as I can, the fundamental features of my proposals. I propose to devote the bulk of this volume to setting forth what can practically be done with one of the most pressing parts of the problem, namely, that relating to those who are out of work, and who, as the result, are more or less destitute. I have many ideas of what might be done with those who are at present cared for in some measure by the State, but I will leave these ideas for the present.

It is not urgent that I should explain how our Poor Law system could be reformed, or what I should like to see done for the Lunatics in Asylums, or the Criminals in Gaols. The persons who are provided for by the State we will, therefore, for the moment, leave out of count. The indoor paupers, the convicts, the inmates of the lunatic asylums are cared for, in a fashion, already. But, over and above all these, there exists some hundreds of thousands who are not quartered on the State, but who are living on the verge of despair, and who at any moment, under circumstances of misfortune, might be compelled to demand relief or support in one shape or another. I will confine myself, therefore, for the present to those who have no helper.

It is possible, I think probable, if the proposals which I am now putting forward are carried out successfully in relation to the lost, homeless, and helpless of the population, that many of those who are at the present moment in somewhat better circumstances will demand that they also shall be allowed to partake in the benefits of the Scheme. But upon this, also, I remain silent. I merely remark that we have, in the recognition of the importance of discipline and organisation, what may be called regimented co-operation, a principle that will be found valuable for solving many social problems other than that of destitution. Of these plans, which are at present being brooded over with a view to their realisation when the time is propitious and the opportunity occurs, I shall have something to say.

What is the outward and visible form of the Problem of the Unemployed? Alas! we are all too familiar with it for any lengthy description to be necessary. The social problem presents itself before us whenever a hungry, dirty and ragged man stands at our door asking if we can give him a crust or a job. That is the social question. What are you to do with that man? He has no money in his pocket, all that he can pawn he has pawned long ago, his stomach is as empty as his purse, and the whole of the clothes upon his back, even if sold on the best terms, would not fetch a shilling. There he stands, your brother, with sixpennyworth of rags to cover his nakedness from his fellow men and not sixpennyworth of victuals within his reach. He asks for work, which he will set to even on his empty stomach and in his ragged uniform, if so be that you will give him something for it, but his hands are idle, for no one employs him. What are you to do with that man? That is the great note of interrogation that confronts Society to-day. Not only in overcrowded England, but in newer countries beyond the sea, where Society has not yet provided a means by which the men can be put upon the land and the land be made to feed the men. To deal with this man is the Problem of the Unemployed. To deal with him effectively you must deal with him immediately, you must provide him in some way or other at once with food, and shelter, and warmth. Next you must find him something to do, something that will test the reality of his desire to work. This test must be more or less temporary, and should be of such a nature as to prepare him for making a permanent livelihood. Then, having trained him, you must provide him where-withal to start life afresh. All these things I propose to do. My Scheme divides itself into three sections, each of which is indispensable for the success of the whole. In this three-fold organisation lies the open secret of the solution of the Social Problem.

The Scheme I have to offer consists in the formation of these people into self-helping and self-sustaining communities, each being a kind of co-operative society, or patriarchal family, governed and disciplined on the principles which have already proved so effective in the Salvation Army.

These communities we will call, for want of a better term, Colonies. There will be—

> (1) The City Colony.
> (2) The Farm Colony.
> (3) The Over-Sea Colony

THE CITY COLONY.

By the City Colony is meant the establishment, in the very centre of the ocean of misery of which we have been speaking, of a number of Institutions to act as Harbours of Refuge for all and any who have been shipwrecked in life, character, or circumstances. These Harbours will gather up the poor destitute creatures, supply their immediate pressing necessities, furnish temporary employment, inspire them with hope for the future, and commence at once a course of regeneration by moral and religious influences.

From these Institutions, which are hereafter described, numbers would, after a short time, be floated off to permanent employment, or sent home to friends happy to receive them on hearing of their reformation. All who remain on our hands would, by varied means, be tested as to their sincerity, industry, and honesty, and as soon as satisfaction was created, be passed on to the Colony of the second class.

THE FARM COLONY.

This would consist of a settlement of the Colonists on an estate in the provinces, in the culture of which they would find employment and obtain support. As the race from the Country to the City has been the cause of much of the distress we have to battle with, we propose to find a substantial part of our remedy by transferring these same people back to the country, that is back again to "the Garden!"

Here the process of reformation of character would be carried forward by the same industrial, moral, and religious methods as have already been commenced in the City, especially including those forms of labour and that knowledge of agriculture which, should the Colonist not obtain employment in this country, will qualify him for pursuing his fortunes under more favourable circumstances in some other land.

From the Farm, as from the City, there can be no question that large numbers, resuscitated in health and character, would be restored to friends up and down the country. Some would find employment in their own callings, others would settle in cottages on a small piece of land that we should provide, or on Co-operative Farms which we intend to promote; while the great bulk, after trial and training, would be passed on to the Foreign Settlement, which would constitute our third class, namely The Over-Sea Colony.

THE OVER-SEA COLONY.

All who have given attention to the subject are agreed that in our Colonies in South Africa, Canada, Western Australia and elsewhere, there are millions of acres of useful land to be obtained almost for the asking, capable of supporting our surplus population in health and comfort, were it a thousand times greater than it is. We propose to secure a tract of land in one of these countries, prepare it for settlement, establish in it authority, govern it by equitable laws, assist it in times of necessity, settling it gradually with a prepared people, and so create a home for these destitute multitudes.

The Scheme, in its entirety, may aptly be compared to A Great Machine, foundationed in the lowest slums and purlieus of our great towns and cities, drawing up into its embrace the depraved and destitute of all classes; receiving thieves, harlots, paupers, drunkards, prodigals, all alike, on the simple conditions of their being willing to work and to conform to discipline. Drawing up these poor outcasts, reforming them, and creating in them habits of industry, honesty, and truth; teaching them methods by which alike the bread that perishes and that which endures to Everlasting Life can be won. Forwarding them from the City to the Country, and there continuing the process of regeneration, and then pouring them forth on to the virgin soils that await their coming in other lands, keeping hold of them with a strong government, and yet making them free men and women; and so laying the foundations, perchance, of another Empire to swell to vast proportions in later times. Why not?

CHAPTER II.

TO THE RESCUE !—THE CITY COLONY.

The first section of my Scheme is the establishment of a Receiving House for the Destitute in every great centre of population. We start, let us remember, from the individual, the ragged, hungry, penniless man who confronts us with despairing demands for food, shelter, and work. Now, I have had some two or three years' experience in dealing with this class. I believe, at the present moment, the Salvation Army supplies more food and shelter to the destitute than any other organisation in London, and it is the experience and encouragement which I have gained in the working of these Food and Shelter Depôts which has largely encouraged me to propound this scheme.

Section I.—FOOD AND SHELTER FOR EVERY MAN.

As I rode through Canada and the United States some three years ago, I was greatly impressed with the superabundance of food which I saw at every turn. Oh, how I longed that the poor starving people, and the hungry children of the East of London and of other centres of our destitute populations, should come into the midst of this abundance, but as it appeared impossible for me to take them to it, I secretly resolved that I would endeavour to bring some of it to them. I am thankful to say that I have already been able to do so on a small scale, and hope to accomplish it ere long on a much vaster one.

With this view, the first Cheap Food Depôt was opened in the East of London two and a half years ago. This has been followed by others, and we have now three establishments : others are being arranged for.

Since the commencement in 1888, we have supplied over three and a half million meals.

Some idea can be formed of the extent to which these Food and Shelter Depôts have already struck their roots into the strata of

Society which it is proposed to benefit, by the following figures, which give the quantities of food sold during the year at our Food Depôts.

FOOD SOLD IN DEPÔTS AND SHELTERS DURING 1889.

Article.	Weight.	Measure.	Remarks.
Soup		116,400 gallons	
Bread	192½ tons	106,964 4lb.-loaves	
Tea	2½ ,,	46,980 gallons	
Coffee	15 cwt.	13,949 ,,	
Cocoa	6 tons	29,229 ,,	
Sugar	25 ,,		300 bags
Potatoes	140 ,,		2,800 ,,
Flour	18 ,,		180 sacks
Peaflour	28½ ,,		288 ,,
Oatmeal	3½ ,,		36 ,,
Rice	12 ,,		120 ,,
Beans	12 ,,		240 ,,
Onions and parsnips	12 ,,		240 ,,
Jam	9 ,,		2,880 jars
Marmalade	6 ,,		1,920 ,,
Meat	15 ,,		
Milk		14,300 quarts	

This includes returns from three Food Depôts and five Shelters. I propose to multiply their number, to develop their usefulness, and to make them the threshold of the whole Scheme. Those who have already visited our Depôts will understand exactly what this means. The majority, however, of the readers of these pages have not done so, and for them it is necessary to explain what they are.

At each of our Depôts, which can be seen by anybody that cares to take the trouble to visit them, there are two departments, one dealing with food, the other with shelter. Of these both are worked together and minister to the same individuals. Many come for food who do not come for shelter, although most of those who come for shelter also come for food, which is sold on terms to cover, as nearly as possible, the cost price and working expenses of the establishment. In this our Food Depôts differ from the ordinary soup kitchens.

There is no gratuitous distribution of victuals. The following is our Price List :—

WHAT IS SOLD AT THE FOOD DEPÔTS.

FOR A CHILD.

	d.		d.
Soup	Per Basin $\frac{1}{4}$	Coffee or Cocoa... ... ... per cup $\frac{1}{4}$
„	With Bread $\frac{1}{2}$	„ „ With Bread and Jam $\frac{1}{2}$

FOR ADULTS.

	d.		d.
Soup	Per Basin $\frac{1}{2}$	Baked Jam Roll... ... ... $\frac{1}{2}$
„	With Bread 1	Meat Pudding and Potatoes 3
Potatoes ... ... $\frac{1}{2}$		Corned Beef „ 2
Cabbage ... ... $\frac{1}{2}$		„ Mutton „ 2
Haricot Beans ... ... $\frac{1}{2}$		Coffee ... ...per cup, $\frac{1}{2}$d. ; per mug 1
Boiled Jam Pudding... ... ... $\frac{1}{2}$		Cocoa ... ... „ $\frac{1}{2}$d. „ 1
„ Plum „ ... ... Each 1		Tea ... ... „ $\frac{1}{2}$d. „ 1
Rice „ ... ... $\frac{1}{2}$		Bread & Butter, Jam, or Marmalade
Baked Plum „ ... ... $\frac{1}{2}$		per slice $\frac{1}{2}$

Soup in own Jugs, 1d. per Quart.
Ready at 10 a.m.

A certain discretionary power is vested in the Officers in charge of the Depôt, and they can in very urgent cases give relief, but the rule is for the food to be paid for, and the financial results show that working expenses are just about covered.

These Cheap Food Depôts I have no doubt have been and are of great service to numbers of hungry starving men, women, and children, at the prices just named, which must be within the reach of all, except the absolutely penniless ; but it is the Shelter that I regard as the most useful feature in this part of our undertaking, for if anything is to be done to get hold of those who use the Depôt, some more favourable opportunity must be afforded than is offered by the mere coming into the food store to get, perhaps, only a basin of soup. This part of the Scheme I propose to extend very considerably.

Suppose that you are a casual in the streets of London, homeless, friendless, weary with looking for work all day and finding none. Night comes on. Where are you to go ? You have perhaps only a few coppers, or it may be, a few shillings, left of the rapidly dwindling store of your little capital. You shrink from sleeping in the open air ; you equally shrink from going to the fourpenny Doss-house where, in the midst of strange and ribald company, you may be robbed of the remnant of the money still in your possession. While at a loss as to what to do, someone who sees you suggests that you should go to our Shelter.   You cannot, of course, go to the Casual Ward of the Workhouse as long as you have any money in your possession.   You come along to one of our Shelters.   On entering you pay fourpence, and are free of the establishment for the night.   You can come in early or late.   The company begins to assemble about five o'clock in the afternoon.   In the women's Shelter you find that many come much earlier and sit sewing, reading or chatting in the sparely furnished but well warmed room from the early hours of the afternoon until bedtime.

You come in, and you get a large pot of coffee, tea, or cocoa, and a hunk of bread.   You can go into the wash-house, where you can have a wash with plenty of warm water, and soap and towels free. Then after having washed and eaten you can make yourself comfortable.   You can write letters to your friends, if you have any friends to write to, or you can read, or you can sit quietly and do nothing.   At eight o'clock the Shelter is tolerably full, and then begins what we consider to be the indispensable feature of the whole concern.   Two or three hundred men in the men's Shelter, or as many women in the women's Shelter, are collected together, most of them strange to each other, in a large room. They are all wretchedly poor—what are you to do with them ?   This is what we do with them.

We hold a rousing Salvation meeting.   The Officer in charge of the Depôt, assisted by detachments from the Training Homes, conducts a jovial free-and-easy social evening.   The girls have their banjos and their tambourines, and for a couple of hours you have as lively a meeting as you will find in London.   There is prayer, short and to the point ; there are addresses, some delivered by the leaders of the meeting, but the most of them the testimonies of those who have been saved at previous meetings, and who, rising in their seats, tell their companions their experiences.   Strange experiences they often are of those who have been down in the very bottomless depths of sin and vice and misery, but who have found at last firm footing on which to stand, and who are, as they say in all sincerity, " as happy as the day is long."   There is a joviality and a genuine good feeling at some of these meetings which is refreshing to the soul.   There are all sorts and conditions of men ; casuals, gaol birds, Out-of-Works, who have come there for the first time, and who find men who last week or last month were even as they themselves are now—still poor but rejoicing in a sense of brotherhood and a consciousness of their being no longer outcasts and forlorn in this

G

wide world. There are men who have at last seen revive before them a hope of escaping from that dreadful vortex, into which their sins and misfortunes had drawn them, and being restored to those comforts that they had feared so long were gone for ever; nay, of rising to live a true and Godly life. These tell their mates how this has come about, and urge all who hear them to try for themselves and see whether it is not a good and happy thing to be soundly saved. In the intervals of testimony—and these testimonies, as every one will bear me witness who has ever attended any of our meetings, are not long, sanctimonious lackadaisical speeches, but simple confessions of individual experience—there are bursts of hearty melody. The conductor of the meeting will start up a verse or two of a hymn illustrative of the experiences mentioned by the last speaker, or one of the girls from the Training Home will sing a solo, accompanying herself on her instrument, while all join in a rattling and rollicking chorus.

There is no compulsion upon anyone of our dossers to take part in this meeting; they do not need to come in until it is over; but as a simple matter of fact they do come in. Any night between eight and ten o'clock you will find these people sitting there, listening to the exhortations and taking part in the singing, many of them, no doubt, unsympathetic enough, but nevertheless preferring to be present with the music and the warmth, mildly stirred, if only by curiosity, as the various testimonies are delivered.

Sometimes these testimonies are enough to rouse the most cynical of observers. We had at one of our shelters the captain of an ocean steamer, who had sunk to the depths of destitution through strong drink. He came in there one night utterly desperate and was taken in hand by our people—and with us taking in hand is no mere phrase, for at the close of our meetings our officers go from seat to seat, and if they see anyone who shows signs of being affected by the speeches or the singing, at once sit down beside him and begin to labour with him for the salvation of his soul. By this means they are able to get hold of the men and to know exactly where the difficulty lies, what the trouble is, and if they do nothing else, at least succeed in convincing them that there is someone who cares for their soul and would do what he could to lend them a helping hand.

The captain of whom I was speaking was got hold of in this way. He was deeply impressed, and was induced to abandon once and for all his habits of intemperance. From that meeting he went an altered man. He regained his position in the merchant service, and twelve months afterwards astonished us all by appearing in the uniform of a captain of a large ocean steamer, to testify to those who were there how low he had been, how utterly he had lost all hold on Society and all hope of the future, when, fortunately led to the Shelter, he found friends, counsel, and salvation, and from that time had never rested until he had regained the position which he had forfeited by his intemperance.

The meeting over, the singing girls go back to the Training Home, and the men prepare for bed. Our sleeping arrangements are somewhat primitive ; we do not provide feather beds, and when you go into our dormitories, you will be surprised to find the floor covered by what look like an endless array of packing cases. These are our beds, and each of them forms a cubicle. There is a mattress laid on the floor, and over the mattress a leather apron, which is all the bedclothes that we find it possible to provide. The men undress, each by the side of his packing box, and go to sleep under their leather covering. The dormitory is warmed with hot water pipes to a temperature of 60 degrees, and there has never been any complaint of lack of warmth on the part of those who use the Shelter. The leather can be kept perfectly clean, and the mattresses, covered with American cloth, are carefully inspected every day, so that no stray specimen of vermin may be left in the place. The men turn in about ten o'clock and sleep until six. We have never any disturbances of any kind in the Shelters. We have provided accommodation now for several thousand of the most helplessly broken-down men in London, criminals many of them, mendicants, tramps, those who are among the filth and offscouring of all things; but such is the influence that is established by the meeting and the moral ascendancy of our officers themselves, that we have never had a fight on the premises, and very seldom do we ever hear an oath or an obscene word. Sometimes there has been trouble outside the Shelter, when men insisted upon coming in drunk or were otherwise violent; but once let them come to the Shelter, and get into the swing of the concern, and we have no trouble with them. In the morning they get up and have their breakfast and, after a short service, go off their various ways.

We find that we can do this, that is to say, we can provide coffee and bread for breakfast and for supper, and a shake-down on the floor in the packing-boxes I have described in a warm dormitory for fourpence a head.

I propose to develop these Shelters, so as to afford every man a locker, in which he could store any little valuables that he may possess. I would also allow him the use of a boiler in the washhouse with a hot drying oven, so that he could wash his shirt over night and have it returned to him dry in the morning. Only those who have had practical experience of the difficulty of seeking for work in London can appreciate the advantages of the opportunity to get your shirt washed in this way—if you have one. In Trafalgar Square, in 1887, there were few things that scandalised the public more than the spectacle of the poor people camped in the Square, washing their shirts in the early morning at the fountains. If you talk to any men who have been on the road for a lengthened period they will tell you that nothing hurts their self-respect more or stands more fatally in the way of their getting a job than the impossibility of getting their little things done up and clean.

In our poor man's "Home" everyone could at least keep himself clean and have a clean shirt to his back, in a plain way, no doubt ; but still not less effective than if he were to be put up at one of the West End hotels, and would be able to secure anyway the necessaries of life while being passed on to something far better. This is the first step.

SOME SHELTER TROPHIES.

Of the practical results which have followed our methods of dealing with the outcasts who take shelter with us we have many striking examples. Here are a few, each of them a transcript of a life experience relating to men who are now active, industrious members of the community upon which but for the agency of these Depôts they would have been preying to this day.

A. S.—Born in Glasgow, 1825. Saved at Clerkenwell, May 19, 1889. Poor parents raised in a Glasgow Slum. Was thrown on the streets at seven years of age, became the companion and associate of thieves, and drifted into crime. The following are his terms of imprisonment :—14 days, 30 days, 30 days, 60 days, 60 days (three times in succession), 4 months, 6 months (twice), 9 months, 18 months, 2 years, 6 years, 7 years (twice), 14 years ; 40 years 3 months and 6 days in the aggregate. Was flogged for violent conduct in gaol 8 times.

W. M. ("Buff").—Born in Deptford, 1864, saved at Clerkenwell, March 31st, 1889. His father was an old Navy man, and earned a decent living as manager. Was sober, respectable, and trustworthy. Mother was a disreputable drunken slattern : a curse and disgrace to husband and family. The home was broken up, and little Buff was given over to the evil influences of his depraved mother. His 7th birthday present from his admiring parent was a "quarten o' gin." He got some education at the One Tun Alley Ragged School, but when nine years old was caught apple stealing, and sent to the Industrial School at Ilford for 7 years. Discharged at the end of his term, he drifted to the streets, the casual wards, and Metropolitan gaols, every one of whose interiors he is familiar with. He became a ringleader of a gang that infested London ; a thorough mendicant and ne'er-do-well; a pest to society. Naturally he is a born leader, and one of those spirits that command a following; consequently, when he got Salvation, the major part of his following came after him to the Shelter, and eventually to God. His character since conversion has been altogether satisfactory, and he is now an Orderly at Whitechapel, and to all appearances a "true lad."

C. W. ("Frisco").—Born in San Francisco, 1862. Saved April 24th, 1889. Taken away from home at the age of eight years, and made his way to Texas. Here he took up life amongst the Ranches as a Cowboy, and varied it with occasional trips to sea, developing into a typical brass and rowdy. He had 2 years for mutiny at sea, 4 years for mule stealing, 5 years for cattle stealing, and has altogether been in gaol for thirteen years and eleven months. He came over to England, got mixed up with thieves and casuals here, and did several short terms of imprisonment. He was met on his release at Millbank by an old chum (Buff) and the Shelter Captain ; came to Shelter, got saved, and has stood firm.

H. A.—Born at Deptford, 1850. Saved at Clerkenwell, January 12th, 1889. Lost mother in early life, step-mother difficulty supervening, and a propensity to misappropriation of small things developed into thieving. He followed the sea, became a hard drinker, a foul-mouthed blasphemer, and a blatant spouter of infidelity. He drifted about for years, ashore and afloat, and eventually reached the Shelter stranded. Here he sought God, and has done well. This summer he had charge of a gang of haymakers sent into the country, and stood the ordeal satisfactorily. He seems honest in his profession, and strives patiently to follow after God. He is at the workshops.

H. S.—Born at A——, in Scotland. Like most Scotch lads although parents were in poor circumstances he managed to get a good education. Early in life he took to newspaper work, and picked up the details of the journalistic profession in several prominent papers in N.B. Eventually he got a position on a provincial newspaper, and having put in a course at Glasgow University, graduated B.A. there. After this he was on the staff of a

Welsh paper. He married a decent girl, and had several little ones, but giving way to drink, lost position, wife, family, and friends. At times he would struggle up and recover himself, and appears generally to have been able to secure a position, but again and again his besetment overcame him, and each time he would drift lower and lower. For a time he was engaged in secretarial work on a prominent London Charity, but fell repeatedly, and at length was dismissed. He came to us an utter outcast, was sent to Shelter and Workshop got saved, and is now in a good situation. He gives every promise, and those best able to judge seem very sanguine that at last a real good work has been accomplished in him.

F. D.—Was born in London, and brought up to the iron trade. Held several good situations, losing one after another, from drink and irregularity. On one occasion, with £20 in his pocket, he started for Manchester, got drunk there, was locked up and fined five shillings, and fifteen shillings costs ; this he paid, and as he was leaving the Court, a gentleman stopped him, saying that he knew his father, and inviting him to his house ; however, with £10 in his pocket, he was too independent, and he declined ; but the gentleman gave him his address, and left him. A few days squandered his cash, and clothes soon followed, all disappearing for drink, and then without a coin he presented himself at the address given to him, at ten o'clock at night. It turned out to be his uncle, who gave him £2 to go back to London, but this too disappeared for liquor. He tramped back to London utterly destitute. Several nights were passed on the Embankment, and on one occasion a gentleman gave him a ticket for the Shelter ; this, however, he sold for 2d. and had a pint of beer, and stopped out all night. But it set him thinking, and he determined next day to raise 4d. and see what a Shelter was like. He came to Whitechapel, became a regular customer, eight months ago got saved, and is now doing well.

F. H.—Was born at Birmingham, 1858. Saved at Whitechapel, March 26th, 1890. Father died in his infancy, mother marrying again. The stepfather was a drunken navvy, and used to knock the mother about, and the lad was left to the streets. At 12 years of age he left home, and tramped to Liverpool, begging his way, and sleeping on the roadsides. In Liverpool he lived about the Docks for some days, sleeping where he could. Police found him and returned him to Birmingham ; his reception being an unmerciful thrashing from the drunken stepfather. He got several jobs as errand-boy, remarkable for his secret pilferings, and two years later left with fifty shillings stolen money, and reached Middlesbrough by road. Got work in a nail factory, stayed nine months, then stole nine shillings from fellow-lodger, and again took the road. He reached Birmingham, and finding a warrant out for him, joined the Navy. He was in the *Impregnable* training-ship three years, behaved himself, only getting "one dozen," and was transferred with character marked "good" to the *Iron Duke* in the China seas; soon got drinking, and was locked up and imprisoned for riotous conduct in almost every port in the stations. He broke ship, and deserted several times, and was a thorough specimen of a bad British tar. He saw gaol in Signapore, Hong Kong, Yokohama, Shanghai, Canton, and other places. In five years returned home, and, after furlough, joined the *Belle Isle* in the Irish station. Whisky here again got hold of him, and excess ruined his constitution. On his leave he had married, and on his discharge joined his wife in Birmingham. For some time he worked as sweeper in the market, but two years ago deserted his wife and family, and came to London, settled down to a loafer's life, lived on the streets with Casual Wards for his home. Eventually came to Whitechapel Shelter, and got saved. He is now a trustworthy, reliable lad; has become reconciled to wife, who came to London to see him, and he bids fair to be a useful man.

J. W. S.—Born in Plymouth. His parents are respectable people. He is clever at his business, and has held good situations. Two years ago he came to London, fell into evil courses, and took to drink. Lost situation after situation, and kept on drinking; lost everything, and came to the streets. He found out Westminster Shelter, and eventually got saved; his parents were communicated with, and help and clothes forthcoming; with Salvation came hope and energy; he got a situation at Lewisham (7d. per hour) at his trade. Four months standing, and is a promising Soldier as well as a respectable mechanic.

J. T.—Born in Ireland; well educated (commercially); clerk and accountant. Early in life joined the Queen's Army, and by good conduct worked his way up. Was orderly-room clerk and paymaster's assistant in his regiment. He led a steady life whilst in the service, and at the expiration of his term passed into the Reserve with a "very good" character. He was a long time unemployed, and this appears to have reduced him to despair, and so to drink. He sank to the lowest ebb, and came to Westminster in a deplorable condition; coatless, hatless, shirtless, dirty altogether, a fearful specimen of what a man of good parentage can be brought to. After being at Shelter some time, he got saved, was passed to Workshops, and gave great satisfaction. At present he is doing clerical work and gives satisfaction as a workman: a good influence in the place.

J. S.—Born in London, of decent parentage. From a child he exhibited thieving propensities; soon got into the hands of the police, and was in and out of gaol continually. He led the life of a confirmed tramp, and roved all over the United Kingdom. He has been in penal servitude three times, and his last term was for seven years, with police supervision. After his release he married a respectable girl, and tried to reform, but circumstances were against him; character he had none, a gaol career only to recommend him, and so he and his wife eventually drifted to destitution. They came to the Shelter, and asked advice ; they were received, and he made application to the sitting Magistrate at Clerkenwell as to a situation, and what he ought to do. The Magistrate helped him, and thanked the Salvation Army for its efforts in behalf of him and such as he, and asked us to look after the applicant. A little work was given him, and after a time a good situation procured. To-day they have a good time ; he is steadily employed, and both are serving God, holding the respect and confidence of neighbours, etc.

E. G.—Came to England in the service of a family of position, and afterwards was butler and upper servant in several houses of the nobility. His health broke down, and for a long time he was altogether unfit for work. He had saved a considerable sum of money, but the cost of doctors and the necessaries of a sick man soon played havoc with his little store, and he became reduced to penury and absolute want. For some time he was in the Workhouse, and, being discharged, he was advised to go to the Shelter. He was low in health as well as in circumstances, and broken in spirit, almost despairing. He was lovingly advised to cast his care upon God, and eventually he was converted. After some time work was obtained as porter in a City warehouse. Assiduity and faithfulness in a year raised him to the position of traveller. To-day he prospers in body and soul, retaining the respect and confidence of all associated with him.

We might multiply these records, but those given show the kind of results attained.

There's no reason to think that influences which have been blessed of God to the salvation of these poor fellows will not be equally efficacious if applied on a wider scale and over a vaster area. The thing to be noted in all these cases is that it was not the mere feeding which effected the result; it was the combination of the feeding with the personal labour for the individual soul. Still, if we had not fed them, we should never have come near enough to gain any hold upon their hearts. If we had merely fed them, they would have gone away next day to resume, with increased energy, the predatory and vagrant life which they had been leading. But when our feeding and Shelter Depôts brought them to close quarters, our officers were literally able to put their arms round their necks and plead with them as brethren who had gone astray. We told them that their sins and sorrows had not shut them out from the love of the Everlasting Father, who had sent us to them to help them with all the power of our strong Organisation, of the Divine authority of which we never feel so sure as when it is going forth to seek and to save the lost.

Section 2.—WORK FOR THE OUT-OF-WORKS.—THE FACTORY.

The foregoing, it will be said, is all very well for your outcast when he has got fourpence in his pocket, but what if he has not got his fourpence? What if you are confronted with a crowd of hungry desperate wretches, without even a penny in their pouch, demanding food and shelter? This objection is natural enough, and has been duly considered from the first.

I propose to establish in connection with every Food and Shelter Depôt a Workshop or Labour Yard, in which any person who comes destitute and starving will be supplied with sufficient work to enable him to earn the fourpence needed for his bed and board. This is a fundamental feature of the Scheme, and one which I think will commend it to all those who are anxious to benefit the poor by enabling them to help themselves without the demoralising intervention of charitable relief.

Let us take our stand for a moment at the door of one of our Shelters. There comes along a grimy, ragged, footsore tramp, his feet bursting out from the sides of his shoes, his clothes all rags, with filthy shirt and towselled hair. He has been, he tells you, on the tramp for the last three weeks, seeking work and finding none, slept last night on the Embankment, and wants to know if you can give him a bite and a sup, and shelter for the night. Has he any money? Not he; he probably spent the last penny he begged or earned in a pipe of tobacco, with which to dull the cravings of his hungry stomach. What are you to do with this man?

Remember this is no fancy sketch—it is a typical case. There are hundreds and thousands of such applicants. Any one who is at all familiar with life in London and our other large towns, will recognise that gaunt figure standing there asking for bread and shelter or for work by which he can obtain both. What can we do with him? Before him Society stands paralysed, quieting its conscience every now and then by an occasional dole of bread and soup, varied with the semi-criminal treatment of the Casual Ward, until the manhood is crushed out of the man and you have in your hands a reckless, despairing, spirit-broken creature, with not even an aspiration to rise above his miserable circumstances, covered with vermin and filth, sinking ever lower and lower, until at last he is hurried out of sight in the rough shell which carries him to a pauper's grave.

I propose to take that man, put a strong arm round him, and extricate him from the mire in which he is all but suffocated. As a first step we will say to him, "You are hungry, here is food; you are homeless, here is a shelter for your head; but remember you must work for your rations. This is not charity; it is work for the workless, help for those who cannot help themselves. There is the labour shed, go and earn your fourpence, and then come in out of the cold and the wet into the warm shelter; here is your mug of coffee and your great chunk of bread, and after you have finished these there is a meeting going on in full swing with its joyful music and hearty human intercourse. There are those who pray for you and with you, and will make you feel yourself a brother among men. There is your shake-down on the floor, where you will have your warm, quiet bed, undisturbed by the ribaldry and curses with which you have been familiar too long. There is the wash-house, where you can have a thorough wash-up at last, after all these days of unwashedness. There is plenty of soap and warm water and clean towels; there, too, you can wash your shirt and have it dried while you sleep. In the morning when you get up there will be breakfast for you, and your shirt will be dry and clean. Then when you are washed and rested, and are no longer faint with hunger, you can go and seek a job, or go back to the Labour shop until something better turns up."

But where and how?

Now let me introduce you to our Labour Yard. Here is no pretence of charity beyond the charity which gives a man remunerative labour. It is not our business to pay men wages. What we propose is to enable those, male or female, who are destitute, to earn their rations and do enough work to pay for their lodging until they are able to go out into the world and earn wages for themselves. There is no compulsion upon any one to resort to our shelter, but if a penniless man wants food he must, as a rule, do work sufficient to pay for what he has of that and of other accommodation. I say as a rule because, of course, our Officers will be allowed to make exceptions in extreme cases, but the rule will be first work then eat.   And that amount of work will be exacted rigorously.   It is that which distinguishes this Scheme from mere charitable relief.

I do not wish to have any hand in establishing a new centre of demoralisation.   I do not want my customers to be pauperised by being treated to anything which they do not earn.   To develop self-respect in the man, to make him feel that at last he has got his foot planted on the first rung of the ladder which leads upwards, is vitally important, and this cannot be done unless the bargain between him and me is strictly carried out.   So much coffee, so much bread, so much shelter, so much warmth and light from me, but so much labour in return from him.

What labour ? it is asked.   For answer to this question I would like to take you down to our Industrial Workshops in Whitechapel. There you will see the Scheme in experimental operation.   What we are doing there we propose to do everywhere up to the extent of the necessity, and there is no reason why we should fail elsewhere if we can succeed there.

Our Industrial Factory at Whitechapel was established this Spring. We opened it on a very small scale.   It has developed until we have nearly ninety men at work.   Some of these are skilled workmen who are engaged in carpentry.   The particular job they have now in hand is the making of benches for the Salvation Army.   Others are engaged in mat making, some are cobblers, others painters, and so forth.   This trial effort has, so far, answered admirably.   No one who is taken on comes for a permanency.   So long as he is willing to work for his rations he is supplied with materials and provided with skilled superintendents.   The hours of work are eight per day.   Here are the rules and regulations under which the work is carried on at present :—

THE SALVATION ARMY SOCIAL REFORM WING.

Temporary Headquarters—

36, UPPER THAMES STREET, LONDON, E.C.

CITY INDUSTRIAL WORKSHOPS.

OBJECTS.—These workshops are open for the relief of the unemployed and destitute, the object being to make it unnecessary for the homeless or workless to be compelled to go to the Workhouse or Casual Ward, food and shelter being provided for them in exchange for work done by them, until they can procure work for themselves, or it can be found for them elsewhere.

PLAN OF OPERATION.—All those applying for assistance will be placed in what is termed the first class. They must be willing to do any kind of work allotted to them. While they remain in the first class, they shall be entitled to three meals a day, and shelter for the night, and will be expected in return to cheerfully perform the work allotted to them.

Promotions will be made from this first-class to the second-class of all those considered eligible by the Labour Directors. They will, in addition to the food and shelter above mentioned, receive sums of money up to 5s. at the end of the week, for the purpose of assisting them to provide themselves with tools, to get work outside.

REGULATIONS.—No smoking, drinking, bad language, or conduct calculated to demoralize will be permitted on the factory premises. No one under the influence of drink will be admitted. Any one refusing to work, or guilty of bad conduct, will be required to leave the premises.

HOURS OF WORK.—7 a.m. to 8.30 a.m.; 9 a.m. to 1 p.m.; 2 p.m. to 5.30 p.m. Doors will be closed 5 minutes after 7, 9, and 2 p.m. Food Checks will be given to all as they pass out at each meal time. Meals and Shelter provided at 272, Whitechapel Road.

Our practical experience shows that we can provide work by which a man can earn his rations. We shall be careful not to sell the goods so manufactured at less than the market prices. In firewood, for instance, we have endeavoured to be rather above the average than below it. As stated elsewhere, we are firmly opposed to injuring one class of workmen while helping another.

Attempts on somewhat similar lines to those now being described have hitherto excited the liveliest feelings of jealousy on the part of the Trade Unions, and representatives of labour. They rightly consider it unfair that labour partly paid for out of the Rates and Taxes, or by Charitable Contributions, should be put upon the market at less than market value, and so compete unjustly with the production of those who have in the first instance to furnish an important quota of the funds by which these Criminal or Pauper workers are supported. No such jealousy can justly exist in relation to our Scheme, seeing that we are endeavouring to raise the standard of labour and are pledged to a war to the death against sweating in every shape and form.

But, it will be asked, how do these Out-of-Works conduct themselves when you get them into the Factory? Upon this point I have a very satisfactory report to render. Many, no doubt, are below par, under-fed, and suffering from ill health, or the consequence of their intemperance. Many also are old men, who have been crowded out of the labour market by their younger generation. But, without making too many allowances on these grounds, I may fairly say that these men have shown themselves not only anxious and willing, but able to work. Our Factory Superintendent reports :—

Of loss of time there has practically been none since the opening, June 29th. Each man during his stay, with hardly an exception, has presented himself punctually at opening time and worked more or less assiduously the whole of the labour hours. The morals of the men have been good, in not more than three instances has there been an overt act of disobedience, insubordination, or mischief. The men, as a whole, are uniformly civil, willing, and satisfied ; they are all fairly industrious, some, and that not a few, are assiduous and energetic. The Foremen have had no serious complaints to make or delinquencies to report.

On the 15th of August I had a return made of the names and trades and mode of employment of the men at work. Of the forty in the shops at that moment, eight were carpenters, twelve labourers, two tailors, two sailors, three clerks, two engineers, while among the rest was a shoemaker, two grocers, a cooper, a sailmaker, a musician, a painter, and a stonemason. Nineteen of these were employed in sawing, cutting and tying up firewood, six were making mats, seven making sacks, and the rest were employed in various odd jobs. Among them was a Russian carpenter who could not speak a word of English. The whole place is a hive of industry which fills the hearts of those who go to see it with hope that something is about to be done to solve the difficulty of the unemployed.

Although our Factories will be permanent institutions they will not be anything more than temporary resting-places to those who avail themselves of their advantages. They are harbours of refuge into which the storm-tossed workman may run and re-fit, so that he may again push out to the ordinary sea of labour and earn his living. The establishment of these Industrial Factories seems to be one of the most obvious duties of those who would effectually deal with the Social Problem. They are as indispensable a link in the chain of deliverance as the Shelters, but they are only a link and not a stopping-place.   And we do not propose that they should be regarded as anything but stepping-stones to better things.

These Shops will also be of service for men and women temporarily unemployed who have families, and who possess some sort of a home. In numerous instances, if by any means these unfortunates could find bread and rent for a few weeks, they would tide over their difficulties, and an untold amount of misery would be averted. In such cases Work would be supplied at their own homes where preferred, especially for the women and children, and such remuneration would be aimed at as would supply the immediate necessities of the hour.   To those who have rent to pay and families to support something beyond rations would be indispensable..

The Labour Shops will enable us to work out our Anti-Sweating experiments.   For instance, we propose at once to commence manufacturing match boxes, for which we shall aim at giving nearly treble the amount at present paid to the poor starving creatures engaged in this work.

In all these workshops our success will depend upon the extent to which we are able to establish and maintain in the minds of the workers sound moral sentiments and to cultivate a spirit of hopefulness and aspiration.   We shall continually seek to impress upon them the fact that while we desire to feed the hungry, and clothe the naked, and provide shelter for the shelterless, we are still more anxious to bring about that regeneration of heart and life which is essential to their future happiness and well-being.

But no compulsion will for a moment be allowed with respect to religion.   The man who professes to love and serve God will be helped because of such profession, and the man who does not will be helped in the hope that he will, sooner or later, in gratitude to God, do the same; but there will be no melancholy misery-making for any.   There is no sanctimonious long face in the Army.   We talk freely about Salvation, because it is to us the very light and joy of our existence.   We are happy, and we wish others to share our joy.   We know by our own experience that life is a very different thing when we have found the peace of God, and are working together with Him for the salvation of the world, instead of toiling for the realisation of worldly ambition or the amassing of earthly gain.

SECTION 3.—THE REGIMENTATION OF THE UNEMPLOYED.

When we have got the homeless, penniless tramp washed, and housed, and fed at the Shelter, and have secured him the means of earning his fourpence by chopping firewood, or making mats or cobbling the shoes of his fellow-labourers at the Factory, we have next to seriously address ourselves to the problem of how to help him to get back into the regular ranks of industry. The Shelter and the Factory are but stepping-stones, which have this advantage, they give us time to look round and to see what there is in a man and what we can make of him.

The first and most obvious thing to do is to ascertain whether there is any demand in the regular market for the labour which is thus thrown upon our hands. In order to ascertain this I have already established a Labour Bureau, the operations of which I shall at once largely extend, at which employers can register their needs, and workmen can register their names and the kind of work they can do.

At present there is no labour exchange in existence in this country. The columns of the daily newspaper are the only substitute for this much needed register. It is one of the many painful consequences arising from the overgrowth of cities. In a village where everybody knows everybody else this necessity does not exist. If a farmer wants a couple of extra men for mowing or some more women for binding at harvest time, he runs over in his mind the names of every available person in the parish. Even in a small town there is little difficulty in knowing who wants employment. But in the cities this knowledge is not available ; hence we constantly hear of persons who would be very glad to employ labour for odd jobs in an occasional stress of work while at the same time hundreds of persons are starving for want of work at another end of the town. To meet this evil the laws of Supply and Demand have created the Sweating

Middlemen, who farm out the unfortunates and charge so heavy a commission for their share that the poor wretches who do the work receive hardly enough to keep body and soul together. I propose to change all this by establishing Registers which will enable us to lay our hands at a moment's notice upon all the unemployed men in a district in any particular trade. In this way we should become the universal intermediary between those who have no employment and those who want workmen.

In this we do not propose to supersede or interfere with the regular Trade Unions. Where Unions exist we should place ourselves in every case in communication with their officials. But the most helpless mass of misery is to be found among the unorganised labourers who have no Union, and who are, therefore, the natural prey of the middleman. Take, for instance, one of the most wretched classes of the community, the poor fellows who perambulate the streets as Sandwich Men. These are farmed out by certain firms. If you wish to send fifty or a hundred men through London carrying boards announcing the excellence of your goods, you go to an advertising firm who will undertake to supply you with as many sandwich men as you want for two shillings or half a crown a day. The men are forthcoming, your goods are advertised, you pay your money, but how much of that goes to the men ? About one shilling, or one shilling and threepence ; the rest goes to the middleman. I propose to supersede this middleman by forming a Co-operative Association of Sandwich Men. At every Shelter there would be a Sandwich Brigade ready in any numbers when wanted. The cost of registration and organisation, which the men would gladly pay, need not certainly amount to more than a penny in the shilling.

All that is needed is to establish a trustworthy and disinterested centre round which the unemployed can group themselves, and which will form the nucleus of a great Co-operative Self-helping Association. The advantages of such a Bureau are obvious. But in this, also, I do not speak from theory. I have behind me the experience of seven months of labour both in England and Australia. In London we have a registration office in Upper Thames Street, where the unemployed come every morning in droves to register their names and to see whether they can obtain situations. In Australia, I see, it was stated in the House of Assembly that our Officers had been instrumental in finding situations for no less than one hundred and thirty-two "Out-of-Works" in a few days. Here, in London, we have succeeded in obtaining employment for a great number, although, of course, it is beyond our power to help all those who apply. We have sent hay-makers down to the country and there is every reason to believe that when our Organisation is better known, and in more extended operation, we shall have a great labour exchange between town and country, so that when there is scarcity in one place and congestion in another, there will be information immediately sent, so that the surplus labour can be drafted into those districts where labour is wanted. For instance, in the harvest seasons, with changeable weather, it is quite a common occurrence for the crops to be seriously damaged for want of labourers, while at the same time there will be thousands wandering about in the big towns and cities seeking work, but finding no one to hire them. Extend this system all over the world, and make it not only applicable to the transfer of workers between the towns and the provinces, but between Country and Country, and it is impossible to exaggerate the enormous advantages which would result. The officer in charge of our experimental Labour Bureau sends me the following notes as to what has already been done through the agency of the Upper Thames Street office :—

SALVATION ARMY SOCIAL REFORM WING.

LABOR BUREAU.

Bureau opened June 16th, 1890. The following are particulars of transactions up to September 26th, 1890 :—

Applications for employment—Men	2462	
Women	208	
		2670
Applications from Employers for Men	128	
„ „ Women	59	
		187
Sent to Work—Men	301	
„ Women	68	
		369
Permanent Situations	146	
Temporary Employment, viz:—Boardmen, Cleaners, &c., &c.	223	
Sent to Workshop in Hanbury Street		165

H

Section 4.—THE HOUSEHOLD SALVAGE BRIGADE.

It is obvious that the moment you begin to find work for the unemployed labour of the community, no matter what you do by way of the registration and bringing together of those who want work and those who want workers, there will still remain a vast residuum of unemployed, and it will be the duty of those who undertake to deal with the question to devise means for securing them employment. Many things are possible when there is a directing intelligence at headquarters and discipline in the rank and file, which would be utterly impossible when everyone is left to go where he pleases, when ten men are running for one man's job, and when no one can be depended upon to be in the way at the time he is wanted. When my Scheme is carried out, there will be in every populous centre a Captain of Industry, an Officer specially charged with the regimentation of unorganised labour, who would be continually on the alert, thinking how best to utilise the waste human material in his district. It is contrary to all previous experience to suppose that the addition of so much trained intelligence will not operate beneficially in securing the disposal of a commodity which is at present a drug in the market.

Robertson, of Brighton, used frequently to remark that every truth was built up of two apparent contradictory propositions. In the same way I may say that the solution of every social difficulty is to be found in the discovery of two corresponding difficulties. It is like the puzzle maps of children. When you are putting one together, you suddenly come upon some awkward piece that will not fit in anywhere, but you do not in disgust and despair break your piece into fragments or throw it away. On the contrary, you keep it by you, knowing that before long you will discover a number of other pieces which it will be impossible to fit in until you fix your unmanageable, unshapely piece in the centre. Now, in the work of piecing together the fragments which lie scattered around the base of our social system we must not despair because we have in the unorganised, untrained labourers that which seems hopelessly out of fit with everything around.   There must be something corresponding to it which is equally useless until he can be brought to bear upon it.   In other words, having got one difficulty in the case of the Out-of-Works, we must cast about to find another difficulty to pair off against it, and then out of two difficulties will arise the solution of the problem.

We shall not have far to seek before we discover in every town and in every country the corresponding element to our unemployed labourer.   We have waste labour on the one hand ; we have waste commodities on the other.   About waste land I shall speak in the next chapter ; I am concerned now solely with waste commodities. Herein we have a means of immediately employing a large number of men under conditions which will enable us to permanently provide for many of those whose hard lot we are now considering.

I propose to establish in every large town what I may call "A Household Salvage Brigade," a civil force of organised collectors, who will patrol the whole town as regularly as the policeman, who will have their appointed beats, and each of whom will be entrusted with the task of collecting the waste of the houses in their circuit. In small towns and villages this is already done, and it will be noticed that most of the suggestions which I have put forth in this book are based upon the central principle, which is that of restoring to the over-grown, and, therefore, uninformed masses of population in our towns the same intelligence and co-operation as to the mutual wants of each and all, that prevails in your small town or village. The latter is the manageable unit, because its dimensions and its needs have not out-grown the range of the individual intelligence and ability of those who dwell therein.   Our troubles in large towns arise chiefly from the fact that the massing of population has caused the physical bulk of Society to outgrow its intelligence. It is as if a human being had suddenly developed fresh limbs which were not connected by any nervous system with the gray matter of his brain.   Such a thing is impossible in the human being, but, unfortunately, it is only too possible in human society.   In the human body no member can suffer without an instantaneous telegram being despatched, as it were, to the seat of intelligence ; the foot or the finger cries out when it suffers, and the whole body suffers with it. So, in a small community, every one, rich and poor, is more or less cognizant of the sufferings of the community. In a large town, where people have ceased to be neighbourly, there is only a congested mass of population settled down on a certain small area without any human ties connecting them together. Here, it is perfectly possible, and it frequently happens, that men actually die of starvation within a few doors of those who, if they had been informed of the actual condition of the sufferer that lay within earshot of their comfortable drawing-rooms, would have been eager to minister the needed relief. What we have to do, therefore, is to grow a new nervous system for the body politic, to create a swift, almost automatic, means of communication between the community as a whole and the meanest of its members, so as to restore to the city what the village possesses.

I do not say that the plan which I have suggested is the only plan or the best plan conceivable. All that I claim for it is that it is the only plan which I can conceive as practicable at the present moment, and that, as a matter of fact, it holds the field alone, for no one, so far as I have been able to discover, even proposes to reconstitute the connection between what I have called the gray matter of the brain of the municipal community and all the individual units which make up the body politic.

Carrying out the same idea I come to the problem of the waste commodities of the towns, and we will take this as an earnest of the working out of the general principle. In the villages there is very little waste. The sewage is applied directly to the land, and so becomes a source of wealth instead of being emptied into great subterranean reservoirs, to generate poisonous gases, which by a most ingenious arrangement, are then poured forth into the very heart of our dwellings, as is the case in the great cities. Neither is there any waste of broken victuals. The villager has his pig or his poultry, or if he has not a pig his neighbour has one, and the collection of broken victuals is conducted as regularly as the delivery of the post. And as it is with broken victuals, so it is with rags and bones, and old iron, and all the *débris* of a household. When I was a boy one of the most familiar figures in the streets of a country town was the man, who, with his small hand-barrow or donkey-cart, made a regular patrol through all the streets once a week, collecting rags, bones, and all other waste materials, buying the same from the juveniles who collected them in specie, not of Her Majesty's current coin, but of common sweetmeats, known as "claggum" or "taffy." When the tootling of his familiar horn was heard the children would bring out their stores, and trade as best they could with the itinerant merchant, with the result that the closets which in our towns to-day have become the receptacles of all kinds of disused lumber were kept then swept and garnished. Now, what I want to know is why can we not establish on a scale commensurate with our extended needs the rag-and-bone industry in all our great towns? That there is sufficient to pay for the collection is, I think, indisputable. If it paid in a small North-country town or Midland village, why would it not pay much better in an area where the houses stand more closely together, and where luxurious living and thriftless habits have so increased that there must be proportionately far more breakage, more waste, and, therefore, more collectable matter than in the rural districts? In looking over the waste of London it has occurred to me that in the *débris* of our households there is sufficient food, if utilised, to feed many of the starving poor, and to employ some thousands of them in its collection, and, in addition, largely to assist the general scheme.

What I propose would be to go to work on something like the following plan :—

London would be divided into districts, beginning with that portion of it most likely to furnish the largest supplies of what would be worth collection. Two men, or a man and a boy, would be told off for this purpose to this district.

Households would be requested to allow a receptacle to be placed in some convenient spot in which the servants could deposit the waste food, and a sack of some description would also be supplied for the paper, rags, &c.

The whole would be collected, say once or twice a week, or more frequently, according to the season and circumstances, and transferred to depôts as central as possible to the different districts.

At present much of this waste is thrown into the dust-bin, there to fester and breed disease. Then there are old newspapers, ragged books, old bottles, tins, canisters, etc. We all know what a number of articles there are which are not quite bad enough to be thrown into the dust heap, and yet are no good to us. We put them on one side, hoping that something may turn up, and as that something very seldom does turn up, there they remain.

Crippled musical instruments, for instance, old toys, broken-down perambulators, old clothes, all the things, in short, for which we have no more need, and for which there is no market within our reach, but which we feel it would be a sin and a shame to destroy.

When I get my Household Salvage Brigade properly organised, beginning, as I said, in some district where we should be likely to meet with most material, our uniformed collectors would call every other day or twice a week with their hand barrow or pony cart. As these men would be under strict discipline, and numbered, the householder would have a security against any abuse of which such regular callers might otherwise be the occasion.

At present the rag and bone man who drives a more or less precarious livelihood by intermittent visits, is looked upon askance by prudent housewives. They fear in many cases he takes the refuse in order to have the opportunity of finding something which may be worth while " picking up," and should he be impudent or negligent there is no authority to whom they can appeal. Under our Brigade, each district would have its numbered officer, who would himself be subordinate to a superior officer, to whom any complaints could be made, and whose duty it would be to see that the officers under his command punctually performed their rounds and discharged their duties without offence.

Here let me disclaim any intention of interfering with the Little Sisters of the Poor, or any other persons, who collect the broken victuals of hotels and other establishments for charitable purposes. My object is not to poach on my neighbour's domains, nor shall I ever be a party to any contentious quarrels for the control of this or that source of supply. All that is already utilised I regard as outside my sphere. The unoccupied wilderness of waste is a wide enough area for the operations of our Brigade. But it will be found in practice that there are no competing agencies. While the broken victuals of certain large hotels are regularly collected, the things before enumerated, and a number of others, are untouched because not sought after.

Of the immense extent to which Food is wasted few people have any notion except those who have made actual experiments. Some years ago, Lady Wolseley established a system of collection from house to house in Mayfair, in order to secure materials for a charitable kitchen which, in concert with Baroness Burdett-Coutts, she had started at Westminster. The amount of the food which she gathered was enormous. Sometimes legs of mutton from which only one or two slices had been cut were thrown into the tub, where they waited for the arrival of the cart on its rounds. It is by no means an excessive estimate to assume that the waste of the kitchens of the West End would provide a sufficient sustenance for all the Out-of-Works who will be employed in our labour sheds at the industrial centres. All that it needs is collection, prompt, systematic, by disciplined men who can be relied upon to discharge their task with punctuality and civility, and whose failure in this duty can be directly brought to the attention of the controlling authority.

Of the utilisation of much of the food which is to be so collected I shall speak hereafter, when I come to describe the second great division of my scheme, namely the Farm Colony. Much of the food collected by the Household Salvage Brigade would not be available for human consumption. In this the greatest care would be exercised, and the remainder would be dispatched, if possible, by barges down the river to the Farm Colony, where we shall meet it hereafter.

But food is only one of the materials which we should handle. At our Whitechapel Factory there is one shoemaker whom we picked off the streets destitute and miserable. He is now saved, and happy, and cobbles away at the shoe leather of his mates. That shoemaker, I foresee, is but the pioneer of a whole army of shoe-makers constantly at work in repairing the cast-off boots and shoes of London. Already in some provincial towns a great business is done by the conversion of old shoes into new. They call the men so employed translators. Boots and shoes, as every wearer of them knows, do not go to pieces all at once or in all parts at once. The sole often wears out utterly, while the upper leather is quite good, or the upper leather bursts while the sole remains practically in a salvable condition ; but your individual pair of shoes and boots are no good to you when any section of them is hopelessly gone to the bad. But give our trained artist in leather and his army of assistants a couple of thousand pairs of boots and shoes, and it will go ill with him if out of the couple of thousand pairs of wrecks he cannot construct five hundred pairs, which, if not quite good, will be immeasurably better than the apologies for boots which cover the feet of many a poor tramp, to say nothing of the thousands of poor children who are at the present moment attending our public schools. In some towns they have already established a Boot and Shoe Fund in order to provide the little ones who come to school with shoes warranted not to let in water between the school house and home. When you remember the 43,000 children who are reported by the School Board to attend the schools of London alone unfed and starving, do you not think there are many thousands to whom we could easily dispose, with advantage, the resurrected shoes of our Boot Factory?

This, however, is only one branch of industry. Take old umbrellas. We all know the itinerant umbrella mender, whose appearance in the neighbourhood of the farmhouse leads the good wife to look after her poultry and to see well to it that the watch-dog is on the premises. But that gentleman is almost the only agency by which old umbrellas can be rescued from the dust heap. Side by side with our Boot Factory we shall have a great umbrella works. The ironwork of one umbrella will be fitted to the stick of another, and even from those that are too hopelessly gone for any further use as umbrellas we shall find plenty of use for their steels and whalebone.

So I might go on. Bottles are a fertile source of minor domestic worry. When you buy a bottle you have to pay a penny for it; but when you have emptied it you cannot get a penny back; no, nor even a farthing. You throw your empty bottle either into the dust heap, or let it lie about. But if we could collect all the waste bottles of London every day, it would go hardly with us if we could not turn a very pretty penny by washing them, sorting them, and sending them out on a new lease of life. The washing of old bottles alone will keep a considerable number of people going.

I can imagine the objection which will be raised by some short-sighted people, that by giving the old, second-hand material a new lease of life it will be said that we shall diminish the demand for new material, and so curtail work and wages at one end while we are endeavouring to piece on something at the other. This objection reminds me of a remark of a North Country pilot who, when speaking of the dulness in the shipbuilding industry, said that nothing would do any good but a series of heavy storms, which would send a goodly number of ocean-going steamers to the bottom, to replace which, this political economist thought, the yards would once more be filled with orders. This, however, is not the way in which work is supplied. Economy is a great auxiliary to trade, inasmuch as the money saved is expended on other products of industry.

There is one material that is continually increasing in quantity, which is the despair of the life of the householder and of the Local Sanitary Authority. I refer to the tins in which provisions are supplied. Nowadays everything comes to us in tins. We have coffee tins, meat tins, salmon tins, and tins *ad nauseam*. Tin is becoming more and more the universal envelope of the rations of man. But when you have extracted the contents of the tin what can you do with it? Huge mountains of empty tins lie about every dustyard, for as yet no man has discovered a means of utilising them when in great masses. Their market price is about four or five shillings a ton, but they are so light that it would take half a dozen trucks to hold a ton. They formerly burnt them for the sake of the solder, but now, by a new process, they are jointed without solder. The problem of the utilisation of the tins is one to which we would have to address ourselves, and I am by no means desponding as to the result.

I see in the old tins of London at least one means of establishing an industry which is at present almost monopolised by our neighbours. Most of the toys which are sold in France on New Year's Day are almost entirely made of sardine tins collected in the French capital. The toy market of England is at present far from being overstocked, for there are multitudes of children who have no toys worth speaking of with which to amuse themselves. In these empty tins I see a means of employing a large number of people in turning out cheap toys which will add a new joy to the households of the poor—the poor to whom every farthing is important, not the rich— the rich can always get toys—but the children of the poor, who live in one room and have nothing to look out upon but the slum or the street. These desolate little things need our toys, and if supplied cheap enough they will take them in sufficient quantities to make it worth while to manufacture them.

A whole book might be written concerning the utilisation of the waste of London. But I am not going to write one. I hope before long to do something much better than write a book, namely, to establish an organisation to utilise the waste, and then if I describe what is being done it will be much better than by now explaining what I propose to do. But there is one more waste material to which it is necessary to allude. I refer to old newspapers and magazines, and books. Newspapers accumulate in our houses until we sometimes burn them in sheer disgust. Magazines and old books lumber our shelves until we hardly know where to turn to put a new volume. My Brigade will relieve the householder from these difficulties, and thereby become a great distributing agency of cheap literature. After the magazine has done its duty in the middle class household it can be passed on to the reading-rooms, work-houses, and hospitals. Every publication issued from the Press that is of the slightest use to men and women will, by our Scheme, acquire a double share of usefulness. It will be read first by its owner, and then by many people who would never otherwise see it.

We shall establish an immense second-hand book shop. All the best books that come into our hands will be exposed for sale, not merely at our central depôts, but on the barrows of our peripatetic colporteurs, who will go from street to street with literature which, I trust, will be somewhat superior to the ordinary pabulum supplied to the poor. After we have sold all we could, and given away all that is needed to public institutions, the remainder will be carried down to our great Paper Mill, of which we shall speak later, in connection with our Farm Colony.

The Household Salvage Brigade will constitute an agency capable of being utilised to any extent for the distribution of parcels newspapers, &c. When once you have your reliable man who will call at every house with the regularity of a postman, and go his beat with the punctuality of a policeman, you can do great things with him. I do not need to elaborate this point. It will be a universal Corps of Commissionaires, created for the service of the public and in the interests of the poor, which will bring us into direct relations with every family in London, and will therefore constitute an unequalled medium for the distribution of advertisements and the collection of information.

It does not require a very fertile imagination to see that when such a house-to-house visitation is regularly established, it will develop in all directions ; and working, as it would, in connection with our Anti-sweating Shops and Industrial Colony, would probably soon become the medium for negotiating sundry household repairs, from a broken window to a damaged stocking. If a porter were wanted to move furniture, or a woman wanted to do charing, or some one to clean windows or any other odd job, the ubiquitous Servant of All who called for the waste, either verbally or by postcard, would re-ceive the order, and whoever was wanted would appear at the time desired without any further trouble on the part of the householder.

One word as to the cost.   There are five hundred thousand houses in the Metropolitan Police district.   To supply every house with a tub and a sack for the reception of waste would involve an initial expenditure which could not possibly be less than one shilling a house. So huge is London, and so enormous the numbers with which we shall have to deal, that this simple preliminary would require a cost of £25,000.   Of course I do not propose to begin on anything like such a vast scale.   That sum, which is only one of the many expenditures involved, will serve to illustrate the extent of the operations which the Household Salvage Brigade will necessitate.   The enterprise is therefore beyond the reach of any but a great and powerful organisation, commanding capital and able to secure loyalty, discipline, and willing service.

CHAPTER III.

TO THE COUNTRY!—THE FARM COLONY.

I leave on one side for a moment various features of the operations which will be indispensable but subsidiary to the City Colony, such as the Rescue Homes for Lost Women, the Retreats for Inebriates, the Homes for Discharged Prisoners, the Enquiry Office for the Discovery of Lost Friends and Relatives, and the Advice Bureau, which will, in time, become an institution that will be invaluable as a poor man's Tribune. All these and other suggestions for saving the lost and helping the poor, although they form essential elements of the City Colony, will be better dealt with after I have explained the relation which the Farm Colony will occupy to the City Colony, and set forth the way in which the former will act as a feeder to the Colony Over Sea.

I have already described how I propose to deal, in the first case, with the mass of surplus labour which will infallibly accumulate on our hands as soon as the Shelters are more extensively established and in good working order. But I fully recognise that when all has been done that can be done in the direction of disposing of the unhired men and women of the town, there will still remain many whom you can neither employ in the Household Salvage Brigade, nor for whom employers, be they registered never so carefully, can be found. What, then, must be done with them? The answer to that question seems to me obvious. They must go upon the land!

The land is the source of all food; only by the application of labour can the land be made fully productive. There is any amount of waste land in the world, not far away in distant Continents, next door to the North Pole, but here at our very doors. Have you ever calculated, for instance, the square miles of unused land which fringe the sides of all our railroads? No doubt some embankments are of material that would baffle the cultivating skill of a Chinese or the careful husbandry of a Swiss mountaineer; but these are exceptions. When other people talk of reclaiming Salisbury Plain, or of cultivating the bare moorlands of the bleak North, I think of the hundreds of square miles of land that lie in long ribbons on the side of each of our railways, upon which, without any cost for cartage, innumerable tons of City manure could be shot down, and the crops of which could be carried at once to the nearest market without any but the initial cost of heaping into convenient trucks. These railway embankments constitute a vast estate, capable of growing fruit enough to supply all the jam that Crosse and Blackwell ever boiled. In almost every county in England are vacant farms, and, in still greater numbers, farms but a quarter cultivated, which only need the application of an industrious population working with due incentive to produce twice, thrice, and four times as much as they yield to-day.

I am aware that there are few subjects upon which there are such fierce controversies as the possibilities of making a livelihood out of small holdings, but Irish cottiers do it, and in regions infinitely worse adapted for the purpose than our Essex corn lands, and possessing none of the advantages which civilization and co-operation place at the command of an intelligently directed body of husbandmen. Talk about the land not being worth cultivating! Go to the Swiss Valleys and examine for yourself the miserable patches of land, hewed out as it were from the heart of the granite mountains, where the cottager grows his crops and makes a livelihood. No doubt he has his Alp, where his cows pasture in summer-time, and his other occupations which enable him to supplement the scanty yield of his farm garden among the crags; but if it pays the Swiss mountaineer in the midst of the eternal snows, far removed from any market, to cultivate such miserable soil in the brief summer of the high Alps, it is impossible to believe that Englishmen, working on English soil, close to our markets and enjoying all the advantages of co-operation, cannot earn their daily bread by their daily toil. The soil of England is not unkindly, and although much is said against our climate, it is, as Mr. Russell Lowell observes, after a lengthened experience of many countries and many climes, " the best climate in the whole world for the labouring man." There are more days in the English year on which a man can work out of doors with a spade with comparative comfort than in any other country under heaven. I do not say that men will make a fortune out of the land, nor do I pretend that we can, under the grey English skies, hope ever to vie with the productiveness of the Jersey farms; but I am prepared to maintain against all comers that it is possible for an industrious man to grow his rations, provided he is given a spade with which to dig and land to dig in. Especially will this be the case with intelligent direction and the advantages of co-operation.

Is it not a reasonable supposition? It always seems to me a strange thing that men should insist that you must first transport your labourer thousands of miles to a desolate, bleak country in order to set him to work to extract a livelihood from the soil when hundreds of thousands of acres lie only half tilled at home or not tilled at all. Is it reasonable to think that you can only begin to make a living out of land when it lies several thousand miles from the nearest market, and thousands of miles from the place where the labourer has to buy his tools and procure all the necessaries of life which are not grown on the spot? If a man can make squatting pay on the prairies or in Australia, where every quarter of grain which he produces has to be dragged by locomotives across the railways of the continent, and then carried by steamers across the wide ocean, can he not equally make the operation at least sufficiently profitable to keep himself alive if you plant him with the same soil within an hour by rail of the greatest markets in the world?

The answer to this is, that you cannot give your man as much soil as he has on the prairies or in the Canadian lumber lands. This, no doubt, is true, but the squatter who settles in the Canadian backwoods does not clear his land all at once. He lives on a small portion of it, and goes on digging and delving little by little, until, after many years of Herculean labour, he hews out for himself, and his children after him, a freehold estate. Freehold estates, I admit, are not to be had for the picking up on English soil, but if a man will but work in England as they work in Canada or in Australia, he will find as little difficulty in making a livelihood here as there.

I may be wrong, but when I travel abroad and see the desperate struggle on the part of peasant proprietors and the small holders in mountainous districts for an additional patch of soil, the idea of cultivating which would make our agricultural labourers turn up their noses in speechless contempt, I cannot but think that our English soil could carry a far greater number of souls to the acre than that which it bears at present. Suppose, for instance, that Essex were suddenly to find itself unmoored from its English anchorage and towed across the Channel to Normandy, or, not to imagine miracles, suppose that an Armada of Chinese were to make a descent on the Isle of Thanet, as did the sea-kings, Hengist and Horsa, does anyone imagine for a moment that Kent, fertile and cultivated as it is, would not be regarded as a very Garden of Eden out of the odd corners of which our yellow-skinned invaders would contrive to extract sufficient to keep themselves in sturdy health? I only suggest the possibility in order to bring out clearly the fact that the difficulty is not in the soil nor in the climate, but in the lack of application of sufficient labour to sufficient land in the truly scientific way.

"What is the scientific way?" I shall be asked impatiently. I am not an agriculturist; I do not dogmatize. I have read much from many pens, and have noted the experiences of many colonies, and I have learned the lesson that it is in the school of practical labour that the most valuable knowledge is to be obtained. Nevertheless, the bulk of my proposals are based upon the experience of many who have devoted their lives to the study of the subject, and have been endorsed by specialists whose experience gives them authority to speak with unquestioning confidence.

Section I.—THE FARM PROPER.

My present idea is to take an estate from five hundred to a thousand acres within reasonable distance of London. It should be of such land as will be suitable for market gardening, while having some clay on it for brick-making and for crops requiring a heavier soil. If possible, it should not only be on a line of railway which is managed by intelligent and progressive directors, but it should have access to the sea and to the river. It should be freehold land, and it should lie at some considerable distance from any town or village. The reason for the latter desideratum is obvious. We must be near London for the sake of our market and for the transmission of the commodities collected by our Household Salvage Brigade, but it must be some little distance from any town or village in order that the Colony may be planted clear out in the open away from the public house, that upas tree of civilisation. A *sine quâ non* of the new Farm Colony is that no intoxicating liquors will be permitted within its confines on any pretext whatever. The doctors will have to prescribe some other stimulant than alcohol for residents in this Colony, But it will be little use excluding alcohol with a strong hand and by cast-iron regulations if the Colonists have only to take a short walk in order to find themselves in the midst of the " Red Lions," and the " Blue Dragons," and the " George the Fourths," which abound in every country town.

Having obtained the land I should proceed to prepare it for the Colonists. This is an operation which is essentially the same in any country. You need water supply, provisions and shelter. All this would be done at first in the simplest possible style. Our pioneer brigade, carefully selected from the competent Out-of-Works in the City Colony, would be sent down to lay out the estate and prepare it for those who would come after. And here let me say that it is a great delusion to imagine that in the riffraff and waste of the labour market there are no workmen to be had except those that are worthless. Worthless under the present conditions, exposed to constant temptations to intemperance no doubt they are, but some of the brightest men in London, with some of the smartest pairs of hands, and the cleverest brains, are at the present moment weltering helplessly in the sludge from which we propose to rescue them.

I am not speaking without book in this matter. Some of my best
Officers to-day have been even such as they. There is an infinite
potentiality of capacity lying latent in our Provincial Tap-rooms
and the City Gin Palaces if you can but get them soundly saved,
and even short of that, if you can place them in conditions where
they would no longer be liable to be sucked back into their old
disastrous habits, you may do great things with them.

I can well imagine the incredulous laughter which will greet my
proposal. "What," it will be said, "do you think that you can
create agricultural pioneers out of the scum of Cockneydom ? " Let
us look for a moment at the ingredients which make up what you
call " the scum of Cockneydom." After careful examination and
close cross-questioning of the Out-of-Works, whom we have already
registered at our Labour Bureau, we find that at least sixty per cent.
are country folk, men, women, boys, and girls, who have left their
homes in the counties to come up to town in the hope of bettering
themselves. They are in no sense of the word Cockneys, and they
represent not the dregs of the country but rather its brighter and
more adventurous spirits who have boldly tried to make their way
in new and uncongenial spheres and have terribly come to grief. Of
thirty cases, selected haphazard, in the various Shelters during the
week ending July 5th, 1890, twenty-two were country-born, sixteen
were men who had come up a long time ago, but did not ever seem
to have settled to regular employ, and four were old military men.
Of sixty cases examined into at the Bureau and Shelters during the
fortnight ending August 2nd, forty-two were country people ; twenty-
six men who had been in London for various periods, ranging from
six months to four years ; nine were lads under eighteen, who had
run away from home and come up to town ; while four were
ex-military. Of eighty-five cases of dossers who were spoken to at
night when they slept in the streets, sixty-three were country people.
A very small proportion of the genuine homeless Out-of-Works are
Londoners bred and born.

There is another element in the matter, the existence of which
will be news to most people, and that is the large proportion of
ex-military men who are among the helpless, hopeless destitute.
Mr. Arnold White, after spending many months in the streets of
London interrogating more than four thousand men whom he found
in the course of one bleak winter sleeping out of doors like animals
returns it as his conviction that at least 20 per cent. are Army

I

Reserve men. Twenty per cent! That is to say one man in every five with whom we shall have to deal has served Her Majesty the Queen under the colours. This is the resource to which these poor fellows come after they have given the prime of their lives to the service of their country. Although this may be largely brought about by their own thriftless and evil conduct, it is a scandal and disgrace which may well make the cheek of the patriot tingle. Still, I see in it a great resource. A man who has been in the Queen's Army is a man who has learnt to obey. He is further a man who has been taught in the roughest of rough schools to be handy and smart, to make the best of the roughest fare, and not to consider himself a martyr if he is sent on a forlorn hope. I often say if we could only get Christians to have one-half of the practical devotion and sense of duty that animates even the commonest Tommy Atkins what a change would be brought about in the world!

Look at poor Tommy! A country lad who gets himself into some scrape, runs away from home, finds himself sinking lower and lower, with no hope of employment, no friends to advise him, and no one to give him a helping hand. In sheer despair he takes the Queen's shilling and enters the ranks. He is handed over to an inexorable drill sergeant, he is compelled to room in barracks where privacy is unknown, to mix with men, many of them vicious, few of them companions whom he would of his own choice select. He gets his rations, and although he is told he will get a shilling a day, there are so many stoppages that he often does not finger a shilling a week. He is drilled and worked and ordered hither and thither as if he were a machine, all of which he takes cheerfully, without even considering that there is any hardship in his lot, plodding on in a dull, stolid kind of way for his Queen and his country, doing his best, also, poor chap, to be proud of his red uniform, and to cultivate his self-respect by reflecting that he is one of the defenders of his native land, one of the heroes upon whose courage and endurance depends the safety of the British realm.

Some fine day at the other end of the world some prancing pro-consul finds it necessary to smash one of the man-slaying machines that loom ominous on his borders, or some savage potentate makes an incursion into territory of a British colony, or some fierce outburst of Mahommedan fanaticism raises up a Mahdi in mid-Africa. In a moment Tommy Atkins is marched off to the troop-ship, and swept across the seas, heart-sick and sea-sick, and miserable exceedingly, to fight the Queen's enemies in foreign parts. When he arrives there he is bundled ashore, brigaded with other troops, marched to the front through the blistering glare of a tropical sun over poisonous marshes in which his comrades sicken and die, until at last he is drawn up in square to receive the charge of tens of thousands of ferocious savages. Far away from all who love him or care for him, foot-sore and travel weary, having eaten perhaps but a piece of dry bread in the last twenty-four hours, he must stand up and kill or be killed. Often he falls beneath the thrust of an assegai or the slashing broadsword of the charging enemy. Then, after the fight is over his comrades turn up the sod where he lies, bundle his poor bones into the shallow pit, and leave him without even a cross to mark his solitary grave. Perhaps he is fortunate and escapes. Yet Tommy goes uncomplainingly through all these hardships and privations, does not think himself a martyr, takes no fine airs about what he has done and suffered, and shrinks uncomplainingly into our Shelters and our Factories, only asking as a benediction from heaven that someone will give him an honest job of work to do. That is the fate of Tommy Atkins. If in our churches and chapels as much as one single individual were to bear and dare, for the benefit of his kind and the salvation of men, what a hundred thousand Tommy Atkins' bear uncomplainingly, taking it all as if it were in the day's work, for their rations and their shilling a day (with stoppages), think you we should not transform the whole face of the world? Yea, verily. We find but very little of such devotion; no, not in Israel.

I look forward to making great use of these Army Reserve men. There are engineers amongst them; there are artillery men and infantry; there are cavalry men, who know what a horse needs to keep him in good health, and men of the transport department, for whom I shall find work enough to do in the transference of the multitudinous waste of London from our town Depôts to the outlying Farm. This, however, is a digression, by the way.

After having got the Farm into some kind of ship-shape, we should select from the City Colonies all those who were likely to be successful as our first settlers. These would consist of men who had been working so many weeks or days in the Labour Factory, or had been under observation for a reasonable time at the Shelters or in the Slums, and who had given evidence of their willingness to work, their amenity to discipline, and their ambition to improve themselves. On arrival at the Farm they would be installed in a barracks, and at once told off to work. In winter time there would be draining, and road-making, and fencing, and many other forms of industry which could go on when the days are short and the nights are long. In Spring, Summertime and Autumn, some would be employed on the land, chiefly in spade husbandry, upon what is called the system of "intensive" agriculture, such as prevails in the suburbs of Paris, where the market gardeners literally create the soil, and which yields much greater results than when you merely scratch the surface with a plough.

Our Farm, I hope, would be as productive as a great market garden There would be a Superintendent on the Colony, who would be a practical gardener, familiar with the best methods of small agriculture, and everything that science and experience shows to be needful for the profitable treatment of the land. Then there would be various other forms of industry continually in progress, so that employment could be furnished, adapted to the capacity and skill of every Colonist. Where farm buildings are wanted, the Colonists must erect them themselves. If they want glass houses, they must put them up. Everything on the Estate must be the production of the Colonists. Take, for instance, the building of cottages. After the first detachment has settled down into its quarters and brought the fields somewhat into cultivation, there will arise a demand for houses. These houses must be built, and the bricks made by the Colonists themselves. All the carpentering and the joinery will be done on the premises, and by this means a sustained demand for work will be created. Then there would be furniture, clothing, and a great many other wants, the supply of the whole of which would create labour which the Colonists must perform.

For a long time to come the Salvation Army will be able to consume all the vegetables and crops which the Colonies will produce. That is one advantage of being connected with so great and growing a concern ; the right hand will help the left, and we shall be able to do many things which those who devote themselves exclusively to colonisation would find it impossible to accomplish. We have seen the large quantities of provisions which are required to supply the Food Depôts in their present dimensions, and with the coming extensions the consumption will be enormously augmented.

On this Farm I propose to carry on every description of "little agriculture."

I have not yet referred to the female side of our operations, but have reserved them for another chapter. It is necessary, however, to bring them in here in order to explain that employment will be created for women as well as men. Fruit farming affords a great opening for female labour, and it will indeed be a change as from Tophet to the Garden of Eden when the poor lost girls on the streets of London exchange the pavements of Piccadilly for the strawberry beds of Essex or Kent.

Not only will vegetables and fruit of every description be raised, but I think that a great deal might be done in the smaller adjuncts of the Farm.

It is quite certain that amongst the mass of people with whom we have to deal there will be a residual remnant of persons to some extent mentally infirm or physically incapacitated from engaging in the harder toils. For these people it is necessary to find work, and I think there would be a good field for their benumbed energies in looking after rabbits, feeding poultry, minding bees, and, in short doing all those little odd jobs about a place which must be attended to, but which will not repay the labour of able-bodied men.

One advantage of the cosmopolitan nature of the Army is that we have Officers in almost every country in the world. When this Scheme is well on the way every Salvation Officer in every land will have it imposed upon him as one of the duties of his calling to keep his eyes open for every useful notion and every conceivable contrivance for increasing the yield of the soil and utilising the employment of waste labour. By this means I hope that there will not be an idea in the world which will not be made available for our Scheme. If an Officer in Sweden can give us practical hints as to how they manage food kitchens for the people, or an Officer in the South of France can explain how the peasants are able to rear eggs and poultry not only for their own use, but so as to be able to export them by the million to England; if a Sergeant in Belgium understands how it is that the rabbit farmers there can feed and fatten and supply our market with millions of rabbits we shall have him over, tap his brains, and set him to work to benefit our people.

By the establishment of this Farm Colony we should create a great school of technical agricultural education. It would be a Working Men's Agricultural University, training people for the life which they would have to lead in the new countries they will go forth to colonise and possess.

Every man who goes to our Farm Colony does so, not to acquire his fortune, but to obtain a knowledge of an occupation and that mastery of his tools which will enable him to play his part in the battle of life. He will be provided with a cheap uniform, which we shall find no difficulty in rigging up from the old clothes of London, and it will go hardly with us, and we shall have worse luck than the ordinary market gardener, if we do not succeed in making sufficient profit to pay all the expenses of the concern, and leave something over for the maintenance of the hopelessly incompetent, and those who, to put it roughly, are not worth their keep.

Every person in the Farm Colony will be taught the elementary lesson of obedience, and will be instructed in the needful arts of husbandry, or some other method of earning his bread. The Agricultural Section will learn the lesson of the seasons and of the best kind of seeds and plants. Those belonging to this Section will learn how to hedge and ditch, how to make roads and build bridges, and generally to subdue the earth and make it yield to him the riches which it never withholds from the industrious and skilful workman. But the Farm Colony, any more than the City Colony, although an abiding institution, will not provide permanently for those with whom we have to deal. It is a Training School for Emigrants, a place where those indispensably practical lessons are given which will enable the Colonists to know their way about and to feel themselves at home wherever there is land to till, stock to rear, and harvests to reap. We shall rely greatly for the peace and prosperity of the Colony upon the sense of brotherhood which will be universal in it from the highest to the lowest. While there will be no systematic wage-paying there will be some sort of rewards and remuneration for honest industry, which will be stored up, for his benefit, as after-wards explained. They will in the main work each for all, and, therefore, the needs of all will be supplied, and any overplus will go to make the bridge over which any poor fellow may escape from the horrible pit and the miry clay from which they themselves have been rescued.

The dulness and deadness of country life, especially in the Colonies, leads many men to prefer a life of hardship and privation in a City slum. But in our Colony they would be near to each other, and would enjoy the advantages of country life and the association and companionship of life in town.

Section 2.—THE INDUSTRIAL VILLAGE.

In describing the operations of the Household Salvage Brigade I have referred to the enormous quantities of good sound food which would be collected from door to door every day of the year. Much of this food would be suitable for human consumption, its waste being next door to sinful. Imagine, for instance, the quantities of soup which might be made from boiling the good fresh meaty bones of the great City! Think of the dainty dishes which a French cook would be able to serve up from the scraps and odds and ends of a single West End kitchen. Good cookery is not an extravagance but an economy, and many a tasty dish is made by our Continental friends out of materials which would be discarded indignantly by the poorest tramp in Whitechapel.

But after all that is done there will remain a mass of food which cannot be eaten by man, but can be converted into food for him by the simple process of passing it through another digestive apparatus. The old bread of London, the soiled, stale crusts can be used in foddering the horses which are employed in collecting the waste. It will help to feed the rabbits, whose hutches will be close by every cottage on the estate, and the hens of the Colony will flourish on the crumbs which fall from the table of Dives. But after the horses and the rabbits and poultry have been served, there will remain a residuum of eatable matter, which can only be profitably disposed of to the voracious and necessary pig. I foresee the rise of a piggery in connection with the new Social Scheme, which will dwarf into insignificance all that exist in Great Britain and Ireland. We have the advantage of the experience of the whole world as to the choice of breeds, the construction of sties, and the rearing of stock. We shall have the major part of our food practically for the cost of collection, and be able to adopt all the latest methods of Chicago for the killing, curing, and disposing of our pork, ham, and bacon.

There are few animals more useful than the pig. He will eat any-thing, live anywhere, and almost every particle of him, from the tip of his nose to the end of his tail, is capable of being converted into a saleable commodity. Your pig also is a great producer of manure, and agriculture is after all largely a matter of manure. Treat the land well and it will treat you well. With our piggery in connection with our Farm Colony there would be no lack of manure.

With the piggery there would grow up a great bacon factory for curing, and that again would make more work. Then as for sausages they would be produced literally by the mile, and all made of the best meat instead of being manufactured out of the very objectionable ingredients too often stowed away in that poor man's favourite ration.

Food, however, is only one of the materials which will be collected by the Household Salvage Brigade. The barges which float down the river with the tide, laden to the brim with the cast-off waste of half a million homes, will bring down an enormous quantity of material which cannot be eaten even by pigs. There will be, for instance, the old bones. At present it pays speculators to go to the prairies of America and gather up the bleached bones of the dead buffaloes, in order to make manure. It pays manu-facturers to bring bones from the end of the earth in order to grind them up for use on our fields. But the waste bones of London; who collects them? I see, as in a vision, barge loads upon barge loads of bones floating down the Thames to the great Bone Factory. Some of the best will yield material for knife handles and buttons, and the numberless articles which will afford ample opportunity in the long winter evenings for the acquisition of skill on the part of our Colonist carvers, while the rest will go straight to the Manure Mill. There will be a constant demand for manure on the part of our ever-increasing nests of new Colonies and our Co-operative Farm, every man in which will be educated in the great doctrine that there is no good agriculture without liberal manuring. And here will be an unfailing source of supply.

Among the material which comes down will be an immense quantity of greasy matter, bits of fat, suet and lard, tallow, strong butter, and all the rancid fat of a great city. For all that we shall have to find use. The best of it will make waggon grease, the rest, after due boiling and straining, will form the nucleus of the raw material which will make our Social Soap a household word throughout the kingdom. After the Manure Works, the Soap Factory will be the natural adjunct of our operations.

The fourth great output of the daily waste of London will be waste paper and rags, which, after being chemically treated, and duly manipulated by machinery, will be re-issued to the world in the shape of paper. The Salvation Army consumes no less than thirty tons of paper every week. Here, therefore, would be one customer for as much paper as the new mill would be able to turn out at the onset ; paper on which we could print the glad tidings of great joy, and tell the poor of all nations the news of salvation for earth and Heaven, full, present, and free to all the children of men.

Then comes the tin. It will go hard with us if we cannot find some way of utilizing these tins, whether we make them into flower-pots with a coat of enamel, or convert them into ornaments, or cut them up for toys or some other purpose. My officers have been instructed to make an exhaustive report on the way the refuse collectors of Paris deal with the sardine tins. The industry of making tin toys will be one which can be practised better in the Farm Colony than in the City. If necessary, we shall bring an accomplished workman from France, who will teach our people the way of dealing with the tin.

In connection with all this it is obvious there would be a constant demand for packing cases, for twine, rope, and for boxes of all kinds ; for carts and cars ; and, in short, we should before long have a complete community practising almost all the trades that are to be found in London, except the keeping of grog shops, the whole being worked upon co-operative principles, but co-operation not for the benefit of the individual co-operator, but for the benefit of the sunken mass that lies behind it.

RULES AND REGULATIONS FOR THE GOVERNMENT OF COLONISTS.

A document containing the Orders and Regulations for the Government of the Colony must be approved and signed by every Colonist before admission. Amongst other things there will be the following :—

1. All Officers must be treated respectfully and implicitly obeyed.

2. The use of intoxicants strictly prohibited, none being allowed within its borders. Any Colonist guilty of violating this Order to be expelled, and that on the first offence.

3. Expulsion for drunkenness, dishonesty, or falsehood will follow the third offence.

4. Profane language strictly forbidden. .

5. No cruelty to be practised on man, woman, child, or animal.

6. Serious offenders against the virtue of women, or of children of either sex, to incur immediate expulsion.

7. After a certain period of probation, and a considerable amount of patience, all who will not work to be expelled.

8. The decision of the Governor of the Colony, whether in the City, or the Farm, or Over the Sea, to be binding in all cases.

9. With respect to penalties, the following rules will be acted upon. The chief reliance for the maintenance of order, as has been observed before, will be placed upon the spirit of love which will prevail throughout the community. But as it cannot be expected to be universally successful, certain penalties will have to be provided :—

(*a*) First offences, except in flagrant cases, will be recorded.

(*b*) The second offence will be published.

(*c*) The third offence will incur expulsion or being handed over to the authorities.

Other regulations will be necessary as the Scheme develops.

There will be no attempt to enforce upon the Colonists the rules and regulations to which Salvation Soldiers are subjected. Those who are soundly saved and who of their own free will desire to become Salvationists will, of course, be subjected to the rules of the Service. But Colonists who are willing to work and obey the orders of the Commanding Officer will only be subject to the foregoing and similar regulations ; in all other things they will be left free.

For instance, there will be no objection to field recreations or any outdoor exercises which conduce to the maintenance of health and spirits. A reading room and a library will be provided, together with a hall, in which they can amuse themselves in the long winter nights and in unfavourable weather. These things are not for the Salvation Army Soldiers, who have other work in the world, but for those who are not in the Army these recreations will be permissible. Gambling and anything of an immoral tendency will be repressed like stealing.

There will probably be an Annual Exhibition of fruit and flowers, at which all the Colonists who have a plot of garden of their own will take part. They will exhibit their fruit and vegetables as well as their rabbits, their poultry and all the other live-stock of the farm.

Every effort will be made to establish village industries, and I am not without hope but that we may be able to restore some of the domestic occupations which steam has compelled us to confine to the great factories. The more the Colony can be made self-supporting the better. And although the hand loom can never compete with Manchester mills, still an occupation which kept the hands of the goodwife busy in the long winter nights, is not to be despised as an element in the economics of the Settlement. While Manchester and Leeds may be able to manufacture common goods much more cheaply than they can be spun at home, even these emporiums, with all their grand improvements in machinery, would be sorely pressed to-day to compete with the hand-loom in many superior classes of work. For instance, we all know the hand-sewn boot still holds its own against the most perfect article that machinery can turn out.

There would be, in the centre of the Colony, a Public Elementary School at which the children would receive training, and side by side with that an Agricultural Industrial School, as elsewhere described.

The religious welfare of the Colony would be looked after by the Salvation Army, but there will be no compulsion to take part in its services. The Sabbath will be strictly observed; no unnecessary work will be done in the Colony on that day, but beyond interdicted labour, the Colonists will be allowed to spend Sunday as they please. It will be the fault of the Salvation Army if they do not find our Sunday Services sufficiently attractive to command their attendance.

Section 3.—AGRICULTURAL VILLAGES.

This brings me to the next feature of the Scheme, the creation of agricultural settlements in the neighbourhood of the Farm, around the original Estate. I hope to obtain land for the purpose of allotments which can be taken up to the extent of so many acres by the more competent Colonists who wish to remain at home instead of going abroad. There will be allotments from three to five acres with a cottage, a cow, and the necessary tools and seed for making the allotment self-supporting. A weekly charge will be imposed for the repayment of the cost of the fixing and stock. The tenant will of course, be entitled to his tenant-right, but adequate precautions will be taken against underletting and other forms by which sweating makes its way into agricultural communities. On entering into possession, the tenant will become responsible for his own and his family's maintenance. I shall stand no longer in the relation of father of the household to him, as I do to the other members of the Colony ; his obligations will cease to me, except in the payment of his rent.

The creation of a large number of Allotment Farms would make the establishment of a creamery necessary, where the milk could be brought in every day and converted into butter by the most modern methods, with the least possible delay. Dairying, which has in some places on the Continent almost developed to a fine art, is in a very backward condition in this country. But by co-operation among the cottiers and an intelligent Headquarter staff much could be done which at present appears impossible.

The tenant will be allowed permanent tenancy on payment of an annual rent or land tax, subject, of course, to such necessary regulations which may be made for the prevention of intemperance and immorality and the preservation of the fundamental features of the Colony. In this way our Farm Colony will throw off small Colonies all round it until the original site is but the centre of a whole series of small farms, where those whom we have rescued and trained will live, if not under their own vine and fig tree, at least in the midst of their own little fruit farm, and surrounded by their small flocks and herds. The cottages will be so many detached residences, each standing in its own ground, not so far away from its neighbours as to deprive its occupants of the benefit of human intercourse.

Section 4.—CO-OPERATIVE FARM.

Side by side with the Farm Colony proper I should propose to renew the experiment of Mr. E. T. Craig, which he found work so successfully at Ralahine. When any members of the original Colony had pulled themselves sufficiently together to desire to begin again on their own account, I should group some of them as partners in a Co-operative Farm, and see whether or no the success achieved in County Clare could not be repeated in Essex or in Kent. I cannot have more unpromising material to deal with than the wild Irishmen on Colonel Vandeleur's estate, and I would certainly take care to be safeguarded against any such mishap as destroyed the early promise of Ralahine.

I shall look upon this as one of the most important experiments of the entire series, and if, as I anticipate, it can be worked successfully, that is, if the results of Ralahine can be secured on a larger scale, I shall consider that the problem of the employment of the people, and the use of the land, and the food supply for the globe, is unquestionably solved, were its inhabitants many times greater in number than they are.

Without saying more, some idea will be obtained as to what I propose from the story of Ralahine related briefly at the close of this volume.

CHAPTER IV.

NEW BRITAIN.—THE COLONY OVER-SEA.

We now come to the third and final stage of the regenerative process. The Colony Over-Sea. To mention Over-Sea is sufficient with some people to damn the Scheme. A prejudice against emigration has been diligently fostered in certain quarters by those who have openly admitted that they did not wish to deplete the ranks of the Army of Discontent at home, for the more discontented people you have here the more trouble you can give the Government, and the more power you have to bring about the general overturn, which is the only thing in which they see any hope for the future. Some again object to emigration on the ground that it is transportation. I confess that I have great sympathy with those who object to emigration as carried on hitherto, and if it be a consolation to any of my critics I may say at once that so far from compulsorily expatriating any Englishman I shall refuse to have any part or lot in emigrating any man or woman who does not voluntarily wish to be sent out.

A journey over sea is a very different thing now to what it was when a voyage to Australia consumed more than six months, when emigrants were crowded by hundreds into sailing ships, and scenes of abominable sin and brutality were the normal incidents of the passage. The world has grown much smaller since the electric telegraph was discovered and side by side with the shrinkage of this planet under the influence of steam and electricity there has come a sense of brotherhood and a consciousness of community of interest and of nationality on the part of the English-speaking people throughout the world. To change from Devon to Australia is not such a change in many respects as merely to cross over from Devon to Normandy. In Australia the Emigrant finds himself among men and women of the same habits, the same language, and in fact the same people, excepting that they live under the southern cross instead of in the northern latitudes. The reduction of the postage between England and the Colonies, a reduction which I hope will soon be followed by the establishment of the Universal Penny Post between the English speaking lands, will further tend to lessen the sense of distance.

The constant travelling of the Colonists backwards and forwards to England makes it absurd to speak of the Colonies as if they were a foreign land. They are simply pieces of Britain distributed about the world, enabling the Britisher to have access to the richest parts of the earth.

Another objection which will be taken to this Scheme is that colonists already over sea will see with infinite alarm the prospect of the transfer of our waste labour to their country. It is easy to understand how this misconception will arise, but there is not much danger of opposition on this score. The working-men who rule the roost at Melbourne object to the introduction of fresh workmen into their labour market, for the same reason that the new Dockers' Union objects to the appearance of new hands at the dock gates, that is for fear the newcomers will enter into unfriendly competition with them. But no Colony, not even the Protectionist and Trade Unionists who govern Victoria, could rationally object to the introduction of trained Colonists planted out upon the land. They would see that these men would become a source of wealth, simply because they would at once become producers as well as consumers, and instead of cutting down wages they would tend directly to improve trade and so increase the employment of the workmen now in the Colony. Emigration as hitherto conducted has been carried out on directly opposite principles to these. Men and women have simply been shot down into countries without any regard to their possession of ability to earn a livelihood, and have consequently become an incubus upon the energies of the community, and a discredit, expense, and burden. The result is that they gravitate to the towns and compete with the colonial workmen, and thereby drive down wages. We shall avoid that mistake. We need not wonder that Australians and other Colonists should object to their countries being converted into a sort of dumping ground, on which to deposit men and women totally unsuited for the new circumstances in which they find themselves.

Moreover, looking at it from the aspect of the class itself, would such emigration be of any enduring value? It is not merely more favourable circumstances that are required by these crowds, but those habits of industry, truthfulness, and self-restraint, which will enable them to profit by better conditions if they could only come to possess them. According to the most reliable information, there are already sadly too many of the same classes we want to help in countries supposed to be the paradise of the working-man.

What could be done with a people whose first enquiry on reaching a foreign land would be for a whisky shop, and who were utterly ignorant of those forms of labour and habits of industry absolutely indispensable to the earning of a subsistence amid the hardships of an Emigrant's life ? Such would naturally shrink from the self-denial the new circumstances inevitably called for, and rather than suffer the inconveniences connected with a settler's life, would probably sink down into helpless despair, or settle in the slums of the first city they came to.

These difficulties, in my estimation, bar the way to the emigration on any considerable scale of the " submerged tenth," and yet I am strongly of opinion, with the majority of those who have thought and written on political economy, that emigration is the only remedy for this mighty evil. Now, the Over-Sea Colony plan, I think, meets these difficulties :—

(1) In the preparation of the Colony for the people.

(2) In the preparation of the people for the Colony.

(3) In the arrangements that are rendered possible for the transport of the people when prepared.

It is proposed to secure a large tract of land in some country suitable to our purpose. We have thought of South Africa, to begin with. We are in no way pledged to this part of the world, or to it alone. There is nothing to prevent our establishing similar settlements in Canada, Australia, or some other land. British Columbia has been strongly urged upon our notice. Indeed, it is certain if this Scheme proves the success we anticipate, the first Colony will be the forerunner of similar communities elsewhere. Africa, however, presents to us great advantages for the moment. There is any amount of land suitable for our purpose which can be obtained, we think, without difficulty. The climate is healthy. Labour is in great demand, so that if by any means work failed on the Colony, there would be abundant opportunities for securing good wages from the neighbouring Companies.

K

Section I.—THE COLONY AND THE COLONISTS.

Before any decision is arrived at, however, information will be obtained as to the position and character of the land; the accessibility of markets for commodities; communication with Europe, and other necessary particulars.

The next business would be to obtain on grant, or otherwise, a sufficient tract of suitable country for the purpose of a Colony, on conditions that would meet its present and future character.

After obtaining a title to the country, the next business will be to effect a settlement in it. This, I suppose, will be accomplished by sending a competent body of men under skilled supervision to fix on a suitable location for the first settlement, erecting such buildings as would be required, enclosing and breaking up the land, putting in first crops, and so storing sufficient supplies of food for the future.

Then a supply of Colonists would be sent out to join them, and from time to time other detachments, as the Colony was prepared to receive them. Further locations could then be chosen, and more country broken up, and before a very long period has passed the Colony would be capable of receiving and absorbing a continuous stream of emigration of considerable proportions.

The next work would be the establishment of a strong and efficient government, prepared to carry out and enforce the same laws and discipline to which the Colonists had been accustomed in England, together with such alterations and additions as the new circumstances would render necessary.

The Colonists would become responsible for all that concerned their own support; that is to say, they would buy and sell, engage in trade, hire servants, and transact all the ordinary business affairs of every-day life.

Our Headquarters in England would represent the Colony in this country on their behalf, and with money supplied by them, when once fairly established, would buy for their agents what they were at the outset unable to produce themselves, such as machinery and the like, also selling their produce to the best advantage.

All land, timber, minerals, and the like, would be rented to the Colonists, all unearned increments, and improvements on the land, would be held on behalf of the entire community, and utilised for its general advantages, a certain percentage being set apart for the extension of its borders, and the continued transmission of Colonists from England in increasing numbers.

Arrangements would be made for the temporary accommodation of new arrivals, Officers being maintained for the purpose of taking them in hand on landing and directing and controlling them generally. So far as possible, they would be introduced to work without any waste of time, situations being ready for them to enter upon ; and any way, their wants would be supplied till this was the case.

There would be friends who would welcome and care for them, not merely on the principle of profit and loss, but on the ground of friendship and religion, many of whom the emigrants would probably have known before in the old country, together with all the social influences, restraints, and religious enjoyments to which the Colonists have been accustomed.

After dealing with the preparation of the Colony for the Colonists, we now come to the preparation of the

COLONISTS FOR THE COLONY OVER-SEA.

They would be prepared by an education in honesty, truth, and industry, without which we could not indulge in any hope of their succeeding. While men and women would be received into the City Colony without character, none would be sent over the sea who had not been proved worthy of this trust.

They would be inspired with an ambition to do well for themselves and their fellow Colonists.

They would be instructed in all that concerned their future career.

They would be taught those industries in which they would be most profitably employed.

They would be inured to the hardships they would have to endure.

They would be accustomed to the economies they would have to practise.

They would be made acquainted with the comrades with whom they would have to live and labour.

They would be accustomed to the Government, Orders, and Regulations which they would have to obey.

They would be educated, so far as the opportunity served, in those habits of patience, forbearance, and affection which would so largely tend to their own welfare, and to the successful carrying out of this part of our Scheme.

TRANSPORT TO THE COLONY OVER-SEA.

We now come to the question of transport. This certainly has an element of difficulty in it, if the remedy is to be applied on a very large scale. But this will appear of less importance if we consider :—

That the largeness of the number will reduce the individual cost. Emigrants can be conveyed to such a location in South Africa, as we have in view, by ones and twos at £8 per head, including land journey ; and, no doubt, were a large number carried, this figure would be reduced considerably.

Many of the Colonists would have friends who would assist them with the cost of passage money and outfit.

All the unmarried will have earned something on the City and Farm Colonies, which will go towards meeting their passage money. In the course of time relatives, who are comfortably settled in the Colony, will save money, and assist their kindred in getting out to them. We have the examples before our eyes in Australia and the United States of how those countries have in this form absorbed from Europe millions of poor struggling people.

All Colonists and emigrants generally will bind themselves in a legal instrument to repay all monies, expenses of passage, outfit, or otherwise, which would in turn be utilised in sending out further contingents.

On the plan named, if prudently carried out, and generously assisted, the transfer of the entire surplus population of this country is not only possible, but would, we think, in process of time, be effected with enormous advantage to the people themselves, to this country, and the country of their adoption. The history of Australia and the United States evidences this. It is quite true the first settlers in the latter were people superior in every way for such an enterprise to the bulk of those we propose to send out. But it is equally true that large numbers of the most ignorant and vicious of our European populations have been pouring into that country ever since without affecting its prosperity, and this Colony Over-Sea would have the immense advantage at the outset which would come from a government and discipline carefully adapted to its peculiar circumstances, and rigidly enforced in every particular.

TRANSPORT TO THE COLONY OVER-SEA.   149

I would guard against misconception in relation to this, Colony Over-Sea by pointing out that all my proposals here are necessarily tentative and experimental.   There is no intention on my part to stick to any of these suggestions if, on maturer consideration and consultation with practical men, they can be improved upon.   Mr. Arnold White, who has already conducted two parties of Colonists to South Africa, is one of the few men in this country who has had practical experience of the actual difficulties of colonisation. I have, through a mutual friend, had the advantage of comparing notes with him very fully, and I venture to believe that there is nothing in this Scheme that is not in harmony with the result of his experience.   In a couple of months this book will be read all over the world.   It will bring me a plentiful crop of suggestions, and, I hope, offers of service from many valuable and experienced Colonists in every country.   In the due order of things the Colony Over-Sea is the last to be started.   Long before our first batch of Colonists is ready to cross the ocean I shall be in a position to correct and revise the proposals of this chapter by the best wisdom and matured experience of the practical men of every Colony in the Empire.

Section 2.—UNIVERSAL EMIGRATION.

We have in our remarks on the Over-Sea Colony referred to the general concensus of opinion on the part of those who have studied the Social Question as to Emigration being the only remedy for the overcrowded population of this country, at the same time showing some of the difficulties which lie in the way of the adoption of the remedy; the dislike of the people to so great a change as is involved in going from one country to another; the cost of their transfer, and their general unfitness for an emigrant's life. These difficulties, as I think we have seen, are fully met by the Over-Sea Colony Scheme. But, apart from those who, driven by their abject poverty, will avail themselves of our Scheme, there are multitudes of people all over the country who would be likely to emigrate could they be assisted in so doing. Those we propose to help in the following manner :—

1. By opening a Bureau in London, and appointing Officers whose business it will be to acquire every kind of information as to suitable countries, their adaptation to, and the openings they present for different trades and callings, the possibility of obtaining land and employment, the rates of remuneration, and the like. These enquiries will include the cost of passage-money, railway fares, outfit, together with every kind of information required by an emigrant.

2. From this Bureau any one may obtain all necessary information.

3. Special terms will be arranged with steamships, railway companies, and land agents, of which emigrants using the Bureau will have the advantage.

4. Introductions will be supplied, as far as possible, to agents and friends in the localities to which the emigrant may be proceeding.

5. Intending emigrants, desirous of saving money, can deposit it through this Bureau in the Army Bank for that purpose.

6. It is expected that government contractors and other employers of labour requiring Colonists of reliable character will apply to this Bureau for such, offering favourable terms with respect to passage-money, employment, and other advantages.

7. No emigrant will be sent out in response to any application from abroad where the emigrant s expenses are defrayed, without references as to character, industry, and fitness.

This Bureau, we think, will be especially useful to women and young girls. There must be a large number of such in this country living in semi-starvation, anyway, with very poor prospects, who would be very welcome abroad, the expense of whose transfer governments, and masters and mistresses alike would be very glad to defray, or assist in defraying, if they could only be assured on both sides of the beneficial character of the arrangements when made.

So widespread now are the operations of the Army, and so extensively will this Bureau multiply its agencies that it will speedily be able to make personal enquiries on both sides, that is in the interest alike of the emigrant and the intended employer in any part of the world.

Section 3.—THE SALVATION SHIP.

When we have selected a party of emigrants whom we believe to be sufficiently prepared to settle on the land which has been got ready for them in the Colony over Sea, it will be no dismal expatriation which will await them. No one who has ever been on the West Coast of Ireland when the emigrants were departing, and has heard the dismal wails which arise from those who are taking leave of each other for the last time on earth, can fail to sympathise with the horror excited in many minds by the very word emigration. But when our party sets out, there will be no violent wrenching of home ties. In our ship we shall export them all—father, mother, and children. The individuals will be grouped in families, and the families will, on the Farm Colony, have been for some months past more or less near neighbours, meeting each other in the field, in the workshops, and in the Religious Services. It will resemble nothing so much as the unmooring of a little piece of England, and towing it across the sea to find a safe anchorage in a sunnier clime. The ship which takes out emigrants will bring back the produce of the farms, and constant travelling to and fro will lead more than ever to the feeling that we and our ocean-sundered brethren are members of one family.

No one who has ever crossed the ocean can have failed to be impressed with the mischief that comes to emigrants when they are on their way to their destination. Many and many a girl has dated her downfall from the temptations which beset her while journeying to a land where she had hoped to find a happier future.

"Satan finds some mischief still for idle hands to do," and he must have his hands full on board an emigrant ship. Look into the steerage at any time, and you will find boredom inexpressible on every face. The men have nothing to do, and an incident of no more importance than the appearance of a sail upon the distant horizon is an event which makes the whole ship talk. I do not see why this should be so. Of course, in the case of conveying passengers and freight, with the utmost possible expedition, for short distances, it would be idle to expect that either time or energies could be spared for the employment or instruction of the passengers. But the case is different when, instead of going to America, the emigrant turns his face to South Africa or remote Australia. Then, even with the fastest steamers, they must remain some weeks or months upon the high seas. The result is that habits of idleness are contracted, bad acquaintances are formed, and very often the moral and religious work of a lifetime is undone.

To avoid these evil consequences, I think we should be compelled to have a ship of our own as soon as possible. A sailing vessel might be found the best adapted for the work. Leaving out the question of time, which would be of very secondary importance with us, the construction of a sailing ship would afford more space for the accommodation of emigrants and for industrial occupation, and would involve considerably less working expenses, besides costing very much less at the onset, even if we did not have one given to us, which I should think would be very probable.

All the emigrants would be under the charge of Army Officers, and instead of the voyage being demoralising, it would be made instructive and profitable. From leaving London to landing at their destination, every colonist would be under watchful oversight, could receive instruction in those particulars where they were still needing it, and be subjected to influences that would be beneficial everyway.

Then we have seen that one of the great difficulties in the direction of emigration is the cost of transport. The expense of conveying a man from England to Australia, occupying as it does some seven or eight weeks, arises not so much from the expense connected with the working of the vessel which carries him, as the amount of provisions he consumes during the passage. Now, with this plan I think that the emigrants might be made to earn at least a portion of this outlay. There is no reason why a man should not work on board ship any more than on land. Of course, nothing much could be done when the weather was very rough; but the average number of days during which it would be impossible for passengers to employ themselves profitably in the time spent between the Channel and Cape Town or Australia would be comparatively few.

When the ship was pitching or rolling, work would be difficult; but even then, when the Colonists get their sea-legs, and are free from the qualmishness which overtakes landsmen when first getting afloat, I cannot see why they should not engage in some form of industrial work far more profitable than yawning and lounging about the deck, to say nothing of the fact that by so doing they would lighten the expense of their transit. The sailors, firemen, engineers, and everybody else connected with a vessel have to work, and there is no reason why our Colonists should not work also.

Of course, this method would require special arrangements in the fitting up of the vessel, which, if it were our own, it would not be difficult to make. At first sight it may seem difficult to find employments on board ship which could be engaged in to advantage, and it might not be found possible to fix up every individual right away; but I think there would be very few of the class and character of people we should take out, with the prior instructions they would have received, who would not have fitted themselves into some useful labour before the voyage ended.

To begin with, there would be a large amount of the ordinary ship's work that the Colonists could perform, such as the preparation of food, serving it out, cleaning the decks and fittings of the ship generally, together with the loading and unloading of cargo. All these operations could be readily done under the direction of permanent hands. Then shoemaking, knitting, sewing, tailoring, and other kindred occupations could be engaged in. I should think sewing-machines could be worked, and, one way or another, any amount of garments could be manufactured, which would find ready and profitable sale on landing, either among the Colonists themselves, or with the people round about.

Not only would the ship thus be a perfect hive of industry, it would also be a floating temple. The Captain, Officers, and every member of the crew would be Salvationists, and all, therefore, alike interested in the enterprise. Moreover, the probabilities are that we should obtain the service of the ship's officers and crew in the most inexpensive manner, in harmony with the usages of the Army everywhere else, men serving from love and not as a mere business. The effect produced by our ship cruising slowly southwards, testifying to the reality of a Salvation for both worlds, calling at all convenient ports, would constitute a new kind of mission work, and drawing out everywhere a large amount of warm practical sympathy. At present the influence of those who go down to the sea in ships is not always in favour of raising the morals and religion of the dwellers in the places where they come. Here, however, would be one ship at least whose appearance foretold no disorder, gave rise to no debauchery, and from whose capacious hull would stream forth an Army of men, who, instead of thronging the grog-shops and other haunts of licentious indulgence, would occupy themselves with explaining and proclaiming the religion of the Love of God and the Brotherhood of Man.

CHAPTER V.

MORE CRUSADES.

I have now sketched out briefly the leading features of the three-fold Scheme by which I think a way can be opened out of " Darkest England," by which its forlorn denizens can escape into the light and freedom of a new life. But it is not enough to make a clear broad road out of the heart of this dense and matted jungle forest; its inhabitants are in many cases so degraded, so hopeless, so utterly desperate that we shall have to do something more than make roads. As we read in the parable, it is often not enough that the feast be prepared, and the guests be bidden; we must needs go into the highways and byways and compel them to come in. So it is not enough to provide our City Colony and our Farm Colony, and then rest on our oars as if we had done our work. That kind of thing will not save the Lost.

It is necessary to organise rescue expeditions to free the miserable wanderers from their captivity, and bring them out into the larger liberty and the fuller life. Talk about Stanley and Emin! There is not one of us but has an Emin somewhere or other in the heart of Darkest England, whom he ought to sally forth to rescue. Our Emins have the Devil for their Mahdi, and when we get to them we find that it is their friends and neighbours who hold them back, and they are, oh, so irresolute! It needs each of us to be as indomitable as Stanley, to burst through all obstacles, to force our way right to the centre of things, and then to labour with the poor prisoner of vice and crime with all our might. But had not the Expeditionary Committee furnished the financial means whereby a road was opened to the sea, both Stanley and Emin would probably have been in the heart of Darkest Africa to this day. This Scheme is our Stanley Expedition. The analogy is very close. I propose to make a road

clear down to the sea.   But alas our poor Emin!   Even when the
road is open, he halts and lingers and doubts.   First he will, and
then he won't, and nothing less than the irresistible pressure of a
friendly and stronger purpose will constrain him to take the road
which has been opened for him at such a cost of blood and treasure.
I now, therefore, proceed to sketch some of the methods by which
we shall attempt to save the lost and to rescue those who are
perishing in the midst of " Darkest England."

Section i.—A SLUM CRUSADE.—OUR SLUM SISTERS.

When Professor Huxley lived as a medical officer in the East of London he acquired a knowledge of the actual condition of the life of many of its populace which led him long afterwards to declare that the surroundings of the savages of New Guinea were much more conducive to the leading of a decent human existence than those in which many of the East-Enders live. Alas, it is not only in London that such lairs exist in which the savages of civilisation lurk and breed. All the great towns in both the Old World and the New have their slums, in which huddle together, in festering and verminous filth, men, women, and children. They correspond to the lepers who thronged the lazar houses of the Middle Ages.

As in those days St. Francis of Assissi and the heroic band of saints who gathered under his orders were wont to go and lodge with the lepers at the city gates, so the devoted souls who have enlisted in the Salvation Army take up their quarters in the heart of the worst slums. But whereas the Friars were men, our Slum Brigade is composed of women. I have a hundred of them under my orders, young women for the most part, quartered all of them in outposts in the heart of the Devil's country. Most of them are the children of the poor who have known hardship from their youth up. Some are ladies born and bred, who have not been afraid to exchange the comfort of a West End drawing-room for service among the vilest of the vile, and a residence in small and fetid rooms whose walls were infested with vermin. They live the life of the Crucified for the sake of the men and women for whom He lived and died. They form one of the branches of the activity of the Army upon which I dwell with deepest sympathy. They are at the front ; they are at close quarters with the enemy.

To the dwellers in decent homes who occupy cushioned pews in fashionable churches there is something strange and quaint in the language they hear read from the Bible, language which habitually refers to the Devil as an actual personality, and to the struggle against sin and uncleanness as if it were a hand to hand death wrestle with the legions of Hell. To our little sisters who dwell in an atmosphere heavy with curses, among people sodden with drink, in quarters where sin and uncleanness are universal, all these Biblical sayings are as real as the quotations of yesterday's price of Consols are to a City man. They dwell in the midst of Hell, and in their daily warfare with a hundred devils it seems incredible to them that anyone can doubt the existence of either one or the other.

The Slum Sister is what her name implies, the Sister of the Slum. They go forth in Apostolic fashion, two-and-two living in a couple of the same kind of dens or rooms as are occupied by the people themselves, differing only in the cleanliness and order, and the few articles of furniture which they contain. Here they live all the year round, visiting the sick, looking after the children, showing the women how to keep themselves and their homes decent, often discharging the sick mother's duties themselves ; cultivating peace, advocating temperance, counselling in temporalities, and ceaselessly preaching the religion of Jesus Christ to the Outcasts of Society.

I do not like to speak of their work. Words fail me, and what I say is so unworthy the theme. I prefer to quote two descriptions by Journalists who have seen these girls at work in the field. The first is taken from a long article which Julia Hayes Percy contributed to the *New York World*, describing a visit paid by her to the slum quarters of the Salvation Army in Cherry Hill Alleys, in the Whitechapel of New York.

Twenty-four hours in the slums—just a night and a day—yet into them were crowded such revelations of misery, depravity, and degradation as having once been gazed upon life can never be the same afterwards. Around and above this blighted neighbourhood flows the tide of active, prosperous life. Men and women travel past in street cars by the Elevated Railroad and across the bridge, and take no thought of its wretchedness, of the criminals bred there, and of the disease engendered by its foulness. It is a fearful menace to the public health, both moral and physical, yet the multitude is as heedless of danger as the peasant who makes his house and plants green vineyards and olives above Vesuvian fires. We are almost as careless and quite as unknowing as we pass the bridge in the late afternoon.

Our immediate destination is the Salvation Army Barracks in Washington Street, and we are going finally to the Salvation Officers—two young women—who have been dwelling and doing a noble mission work for months in one of the worst corners of New York's most wretched quarter. These Officers are not living under the ægis of the Army, however. The blue bordered flag is furled out of sight, the uniforms and poke bonnets are laid away, and there are no drums or tambourines. "The banner over them is love" of their fellow-creatures among whom they dwell upon an equal plane of poverty, wearing no better clothes than the rest, eating coarse and scanty food, and sleeping upon hard cots or upon the floor. Their lives are consecrated to God's service among the poor of the earth. One is a woman in the early prime of vigorous life, the other a girl of eighteen. The elder of these devoted women is awaiting us at the barracks to be our guide to Slumdom. She is tall, slender, and clad in a coarse brown gown, mended with patches. A big gingham apron, artistically rent in several places, is tied about her waist. She wears on old plaid woollen shawl and an ancient brown straw hat. Her dress indicates extreme poverty, her face denotes perfect peace. "This is Em," says Mrs. Ballington Booth, and after this introduction we sally forth.

More and more wretched grows the district as we penetrate further. Em pauses before a dirty, broken, smoke-dimmed window, through which in a dingy room are seen a party of roughs, dark-looking men, drinking and squabbling at a table. "They are our neighbours in the front." We enter the hall-way and proceed to the rear room. It is tiny, but clean and warm. A fire burns on the little cracked stove, which stands up bravely on three legs, with a brick eking out its support at the fourth corner. A tin lamp stands on the table, half-a-dozen chairs, one of which has arms, but must have renounced its rockers long ago, and a packing box, upon which we deposit our shawls, constitute the furniture. Opening from this is a small dark bedroom, with one cot made up and another folded against the wall. Against a door, which must communicate with the front room, in which we saw the disagreeable-looking men sitting, is a wooden table for the hand-basin. A small trunk and a barrel of clothing complete the inventory.

Em's sister in the slum work gives us a sweet shy welcome. She is a Swedish girl, with the fair complexion and crisp, bright hair peculiar to the Scandinavian blonde-type. Her head reminds me of a Grenze that hangs in the Louvre, with its low knot of rippling hair, which fluffs out from her brow and frames a dear little face with soft childish outlines, a nez retrousse, a tiny mouth, like a crushed pink rose, and wistful blue eyes. This girl has been a Salvationist for two years. During that time she has learned to speak, read, and write English, while she has constantly laboured among the poor and wretched.

The house where we find ourselves was formerly notorious as one of the worst in the Cherry Hill district. It has been the scene of some memorable crimes, and among them that of the Chinaman who slew his Irish wife, after the manner of "Jack the Ripper," on the staircase leading to the second floor. A notable change has taken place in the tenement since Mattie and Em have lived there, and their gentle influence is making itself felt in the neighbouring houses as well. It is nearly eight o'clock when we sally forth. Each of us carries a handful of printed slips bearing a text of Scripture and a few words of warning to lead the better life.

"These furnish an excuse for entering places where otherwise we could not go," explains Em.

After arranging a rendezvous, we separate. Mattie and Liz go off in one direction, and Em and I in another. From this our progress seems like a descent into Tartarus. Em pauses before a miserable-looking saloon, pushes open the low, swinging door, and we go in. It is a low-ceiled room, dingy with dirt, dim with the smoke, nauseating with the fumes of sour beer and vile liquor. A sloppy bar extends along one side, and opposite is a long table, with indescribable viands littered over it, interspersed with empty glasses, battered hats, and cigar stumps. A motley crowd of men and women jostle in the narrow space. Em speaks to the soberest looking of the lot. He listens to her words, others crowd about. Many accept the slips we offer, and gradually, as the throng separates to make way, we gain the further end of the apartment. Em's serious, sweet, saint-like face I follow like a star. All sense of fear slips from me, and a great pity fills my soul as I look upon the various types of wretchedness.

As the night wears on, the whole apartment seems to wake up. Every house is alight; the narrow sidewalks and filthy streets are full of people. Miserable little children, with sin-stamped faces, dart about like rats; little ones who ought to be in their cribs shift for themselves, and sleep on cellar doors and areas, and under carts; a few vendors are abroad with their wares, but the most of the traffic going on is of a different description. Along Water Street are women conspicuously dressed in gaudy colours. Their heavily-painted faces are bloated or pinched; they shiver in the raw night air. Liz speaks to one, who replies that she would like to talk, but dare not, and as she says this an old hag comes to the door and cries :—

"Get along; don't hinder her work!

During the evening a man to whom Em has been talking has told her :—

"You ought to join the Salvation Army; they are the only good women who bother us down here. I don't want to lead that sort of life; but I must go where it is light and warm and clean after working all day, and there isn't any place but this to come to " exclaimed the man.

"You will appreciate the plea to-morrow when you see how the people live," Em says, as we turn our steps toward the tenement room, which seems like an

L

oasis of peace and purity after the howling desert we have been wandering in.
Em and Mattie brew some oatmeal gruel, and being chilled and faint we enjoyed a cup of it.   Liz and I share a cot in the outer room.   We are just going
to sleep when agonised cries ring out through the night ; then the tones of a
woman's voice pleading pitifully reach our ears.   We are unable to distinguish
her words, but the sound is heart-rending.   It comes from one of those dreadful
Water Street houses, and we all feel that a tragedy is taking place.   There is a
sound of crashing blows and then silence.

It is customary in the slums to leave the house door open perpetually, which
is convenient for tramps, who creep into the hall-ways to sleep at night, thereby
saving the few pence it costs to occupy a "spot" in the cheap lodging houses.
Em and Mat keep the corridor without their room beautifully clean, and so it has
become an especial favourite stamping ground for these vagrants.   We were told
this when Mattie locked and bolted the door and then tied the keys and the door-
handle together.   So we understand why there are shuffling steps along the
corridor, bumping against the panels of the door, and heavily breathing without
during the long hours of the night.

All day Em and Mat have been toiling among their neighbours, and the night
before last they sat up with a dying woman.   They are worn out and sleep
heavily.   Liz and I lie awake and wait for the coming of the morning ; we are
too oppressed by what we have seen and heard to talk.

In the morning Liz and I peep over into the rear houses where we heard
those dreadful shrieks in the night.   There is no sign of life, but we discover
enough filth to breed diphtheria and typhoid throughout a large section.   In the
area below our window there are several inches of stagnant water, in which is
heaped a mass of old shoes, cabbage heads, garbage, rotten wood, bones, rags
and refuse, and a few dead rats.   We understand now why Em keeps her room
ull of disinfectants.   She tells us that she dare not make any appeal to the
sanitary authorities, either on behalf of their own or any other dwelling, for fear
of antagonizing the people, who consider such officials as their natural enemies.

The first visit we pay is up a number of eccentric little flights of shaky steps
interspersed with twists of passageway.   The floor is full of holes.   The stairs
have been patched here and there, but look perilous and sway beneath the feet,
A low door on the landing is opened by a bundle of rags and filth, out of which
issues a woman's voice in husky tones, bidding us enter.   She has *La grippe*.
We have to stand very close together, for the room is small, and already
contains three women, a man, a baby, a bedstead, a stove, and indescribable
dirt.   The atmosphere is rank with impurity.   The man is evidently dying.
Seven weeks ago he was "gripped."   He is now in the last stages of pneumonia.
Em has tried to induce him to be removed to the hospital, and he gasps out his
desire "to die in comfort in my own bed."   Comfort !   The "bed" is a rack
heaped with rags.   Sheets, pillow-cases, and night-clothes are not in vogue in

the slums. A woman lies asleep on the dirty floor with her head under the table. Another woman, who has been sharing the night watch with the invalid's wife, is finishing her morning meal, in which roast oysters on the half shell are conspicuous. A child that appears never to have been washed toddles about the floor and tumbles over the sleeping woman's form. Em gives it some gruel, and ascertains that its name is " Christine."

The dirt, crowding, and smells in the first place are characteristic ot half a dozen others we visited. We penetrate to garrets and descend into cellars. The "rear houses" are particularly dreadful. Everywhere there is decaying garbage lying about, and the dead cats and rats are evidence that there are mighty hunters among the gamins of the Fourth Ward. We find a number ill from the grip and consequent maladies. None of the sufferers will entertain the thought of seeking a hospital. One probably voices the opinion of the majority when he declares that "they'll wash you to death there." For these people a bath possesses more terror than the gallows or the grave.

In one room, with a wee window, lies a woman dying of consumption ; wasted, wan, and wretched, lying on rags and swarming with vermin. Her little son, a boy of eight years, nestles beside her. His cheeks are scarlet, his eyes feverishly bright, and he has a hard cough.

"It's the chills, mum," says the little chap.

Six beds stand close together in another room ; one is empty. Three days ago a woman died there and the body has just been taken away. It hasn't disturbed the rest of the inmates to have death present there. A woman is lying on the wrecks of a bedstead, slats and posts sticking out in every direction from the rags on which she reposes.

"It broke under me in the night," she explains. A woman is sick and wants Liz to say a prayer. We kneel on the filthy floor. Soon all my faculties are absorbed in speculating which will arrive first, the "Amen" or the "B flat" which is wending its way towards me. This time the bug does not get there, and I enjoy grinding him under the sole of my Slum shoe when the prayer is ended.

In another room we find what looks like a corpse. It is a woman in an opium stupor. Drunken men are brawling around her.

Returning to our tenement, Em and Liz meet us, and we return to our experi‐ ence. The minor details vary slightly, but the story is the same piteous tale of woe everywhere, and crime abounding, conditions which only change to a prison, a plunge in the river, or the Potter's field.

The Dark Continent can show no lower depth ot degradation than that sounded by the dwellers of the dark alleys in Cherry Hill. There isn't a vice missing in that quarter. Every sin in the Decalogue flourishes in that feeder of penitentiaries and prisons. And even as its moral foulness permeates and poisons the veins of our social life so the malarial filth with which the locality reeks must sooner or later spread disease and death.

An awful picture, truly, but one which is to me irradiated with the love-light which shone in the eyes of "Em's serious, sweet, saint-like face."

Here is my second. It was written by a Journalist who had just witnessed the scene in Whitechapel. He writes :—

I had just passed Mr. Barnett's church when I was stopped by a small crowd at a street corner. There were about thirty or forty men, women, and children standing loosely together, some others were lounging on the opposite side of the street round the door of a public-house. In the centre of the crowd was a plain-looking little woman in Salvation Army uniform, with her eyes closed, praying the "dear Lord that he would bless these dear people, and save them, save them now!" Moved by curiosity, I pressed through the outer fringe of the crowd, and in doing so, I noticed a woman of another kind, also invoking Heaven, but in an altogether different fashion. Two dirty tramp-like men were listening to the prayer, standing the while smoking their short cutty pipes. For some reason or other they had offended the woman, and she was giving them a piece of her mind. They stood stolidly silent while she went at them like a fiend. She had been good-looking once, but was now horribly bloated with drink, and excited by passion. I heard both voices at the same time. What a contrast! The prayer was over now, and a pleading earnest address was being delivered.

"You are wrong," said the voice in the centre "you know you are; all this misery and poverty is a proof of it. You are prodigals. You have got away from your Father's house, and you are rebelling against Him every day. Can you wonder that there is so much hunger, and oppression, and wretchedness allowed to come upon you? In the midst of it all your Father loves you. He wants you to return to Him; to turn your backs upon your sins; abandon your evil doings; give up the drink and the service of the devil. He has given His Son Jesus Christ to die for you. He wants to save you. Come to His feet. He is waiting. His arms are open. I know the devil has got fast hold of you; but Jesus will give you grace to conquer him. He will help you to master your wicked habits and your love of drink. But come to Him now. God is love. He loves me. He loves you. He loves us all. He wants to save us all."

Clear and strong the voice, eloquent with the fervour of intense feeling, rang through the little crowd, past which streamed the ever-flowing tide of East End life. And at the same time that I heard this pure and passionate invocation to love God and be true to man I heard a voice on the outskirts, and it said this: "You ——— swine! I'll knock the vitals out of yer. None of your ——— impudence to me. ———

your ———— eyes, what do you mean by telling me that ? You know what you ha' done, and now you are going to the Salvation Army. I'll let them know you, you dirty rascal." The man shifted his pipe. " What's the matter?" " Matter !" screamed the virago hoarsely. " ———— yer life, don't you know what's the matter ? I'll matter ye, you ———— hound. By God ! I will, as sure as I'm alive. Matter ! you know what's the matter." And so she went on, the men standing silently smoking until at last she took herself off, her mouth full of oaths and cursing, to the public-house. It seemed as though the presence, and spirit, and words of the Officer, who still went on with the message of mercy, had some strange effect upon them, which made these poor wretches impervious to the taunting, bitter sarcasms of this brazen, blatant virago.

" God is love." Was it not, then, the accents of God's voice that sounded there above the din of the street and the swearing of the slums? Yea, verily, and that voice ceases not and will not cease, so long as the Slum Sisters fight under the banner of the Salvation Army.

To form an idea of the immense amount of good, temporal and spiritual, which the Slum Sister is doing ; you need to follow them into the kennels where they live, preaching the Gospel with the mop and the scrubbing brush, and driving out the devil with soap and water. In one of our Slum posts, where the Officer's rooms were on the ground floor, about fourteen other families lived in the same house. One little water-closet in the back yard had to do service for the whole place. As for the dirt, one Officer writes, " It is impossible to scrub the Homes ; some of them are in such a filthy condition. When they have a fire the ashes are left to accumulate for days. The table is very seldom, if ever, properly cleaned, dirty cups and saucers lie about it, together with bits of bread, and if they have bloaters the bones and heads are left on the table. Sometimes there are pieces of onions mixed up with the rest. The floors are in a very much worse condition than the street pavements, and when they are supposed to clean them they do it with about a pint of dirty water. When they wash, which is rarely, for washing to them seems an unnecessary work, they do it in a quart or two of water, and sometimes boil the things in some old saucepan in which they cook their food. They do this simply because they have no larger vessel to wash in. The vermin fall off the walls and ceiling on you while you are standing in the rooms. Some of the walls are covered with marks where they have killed them. Many people in the summer sit on the door steps all night, the reason for this being, that their rooms are so close from the heat and so unendurable from the vermin that they prefer staying out in the cool night air. But as they cannot stay anywhere long without drinking, they send for beer from the neighbouring public—alas ! never far away—and pass it from one doorway to another, the result being singing, shouting and fighting up till three and four o'clock in the morning."

I could fill volumes with stories of the war against vermin, which is part of this campaign in the slums, but the subject is too revolting to those who are often indifferent to the agonies their fellow creatures suffer, so long as their sensitive ears are not shocked by the mention of so painful a subject. Here, for instance, is a sample of the kind of region in which the Slum Sisters spend themselves :—

"In an apparently respectable street near Oxford street, the Officers where visiting one day when they saw a very dark staircase leading into a cellar, and thinking it possible that someone might be there they attempted to go down, and yet the staircase was so dark they thought it impossible for anyone to be there. However, they tried again and groped their way along in the dark for some time until at last they found the door and entered the room. At first they could not discern anything because of the darkness. But after they got used to it they saw a filthy room. There was no fire in the grate, but the fire-place was heaped up with ashes, an accumulation of several weeks at least. At one end of the room there was an old sack of rags and bones partly emptied upon the floor, from which there came a most unpleasant odour. At the other end lay an old man very ill. The apology for a bed on which he lay was filthy and had neither sheets nor blankets. His covering consisted of old rags. His poor wife, who attended on him, appeared to be a stranger to soap and water. These Slum Sisters nursed the old people, and on one occasion undertook to do their washing, and they brought it home to their copper for this purpose, but it was so infested with vermin that they did not know how to wash it. Their landlady, who happened to see them, forbade them ever to bring such stuff there any more. The old man, when well enough, worked at his trade, which was tailoring. They had two shillings and sixpence per week from the parish."

Here is a report from the headquarters of our Slum Brigade as to the work which the Slum Sisters have done.

It is almost four years since the Slum Work was started in London. The principal work done by our first Officers was that of visiting the sick, cleansing the homes of the Slummers, and of feeding the hungry. The following are a few of the cases of those who have gained temporally, as well as spiritually, through our work :—

Mrs. W.—Of Haggerston Slum. Heavy drinker, wrecked home, husband a drunkard, place dirty and filthy, terribly poor. Saved now over two years, home AI., plenty of employment at cane-chair bottoming; husband now saved also.

Mrs. R.—Drury Lane Slum. Husband and wife, drunkards; husband very lazy, only worked when starved into it. We found them both out of work, home furnitureless, in debt. She got saved, and our lasses prayed for him to get work. He did so, and went to it. He fell out again a few weeks after, and beat his wife. She sought employment at charing and office cleaning, got it, and has been regularly at work since. He too got work. He is now a teetotaler. The home is very comfortable now, and they are putting money in the bank.

A. M. in the Dials. Was a great drunkard, thriftless, did not go to the trouble of seeking work. Was in a Slum meeting, heard the Captain speak on " Seek first the Kingdom of God ! " called out and said, " Do you mean that if I ask God for work, He will give it me ? " Of course she said, " Yes." He was converted that night, found work, and is now employed in the Gas Works, Old Kent Road.

Jimmy is a soldier in the Boro' Slum. Was starving when he got converted through being out of work. Through joining the Army, he was turned out of his home. He found work, and now owns a coffee-stall in Billingsgate Market, and is doing well.

Sergeant R.—Of Marylebone Slum. Used to drink, lived in a wretched place in the famous Charles Street, had work at two places, at one of which he got 5s. a week, and the other 10s., when he got saved; this was starvation wages, on which to keep himself, his wife, and four children. At the 10s. a week work he had to deliver drink for a spirit merchant; feeling condemned over it, he gave it up, and was out of work for weeks. The brokers were put in, but the Lord rescued him just in time. The 5s. a week employer took him afterwards at 18s., and he is now earning 22s., and has left the ground-floor Slum tenement for a better house.

H.—Nine Elms Slum. Was saved on Easter Monday, out of work several weeks before, is a labourer, seems very earnest, in terrible distress. We allow his wife 2s. 6d. a week for cleaning the hall (to help them). In addition to that, she gets another 2s. 6d. for nursing, and on that husband, wife, and a couple of children pay the rent of 2s. a week and drag out an existence. I have tried to get work for this man, but have failed.

T.—Of Rotherhithe Slum. Was a great drunkard, is a carpenter; saved about nine months ago, but, having to work in a public-house on a Sunday, he gave it up; he has not been able to get another job, and has nothing but what we have given him for making seats.

Emma Y.—Now a Soldier of the Marylebone Slum Post, was a wild young Slummer when we opened in the Boro'; could be generally seen in the streets, wretchedly clad, her sleeves turned up, idle, only worked occasionally, got saved two years ago, had terrible persecution in her home. We got her a situation, where she has been for nearly eighteen months, and is now a good servant.

Lodging-House Frank.—At twenty-one came into the possession of £750, but, through drink and gambling, lost it all in six or eight months, and for over seven years he has tramped about from Portsmouth, through the South of England, and South Wales, from one lodging-house to another, often starving, drinking when he could get any money; thriftless, idle, no heart for work. We found him in a lodging-house six months ago, living with a fallen girl; got them both saved and married; five weeks after he got work as a carpenter at 30s. a week. He has a home of his own now, and promises well to make an Officer.

The Officer who furnishes the above reports goes on to say :—

I can't call the wretched dwelling home, to which drink had brought Brother and Sister X. From a life of luxury, they drifted down by degrees to one room in a Slum tenement, surrounded by drunkards and the vilest characters. Their lovely half-starved children were compelled to listen to the foulest language, and hear fighting and quarrelling, and alas, alas, not only to hear it in the adjoining rooms, but witness it within their own. For over two years they have been delivered from the power of the cursed drink. The old rookery is gone, and now they have a comfortably-furnished home. Their children give evidence of being truly converted, and have a lively gratitude for their father's salvation. One boy of eight said, last Christmas Day, "I remember when we had only dry bread for Christmas; but to-day we had a goose and two plum-puddings." Brother X. was dismissed in disgrace from his situation as commercial traveller before his conversion; to-day he is chief man, next to his employer, in a large business house.

He says :—

I am perfectly satisfied that very few of the lowest strata of Society are unwilling to work if they could get it. The wretched hand-to-mouth existence many of them have to live disheartens them, and makes life with them either a feast or a famine, and drives those who have brains enough to crime.

The results of our work in the Slums may be put down as :—

1st. A marked improvement in the cleanliness of the homes and children ; disappearance of vermin, and a considerable lessening of drunkenness.

2nd. A greater respect for true religion, and especially that of the Salvation Army.

3rd. A much larger amount of work is being done now than before our going there.

4th. The rescue of many fallen girls.

5th. The Shelter work seems to us a development of the Slum work.

In connection with our Scheme, we propose to immediately increase the numbers of these Slum Sisters, and to add to their usefulness by directly connecting their operations with the Colony, enabling them thereby to help the poor people to conditions of life more favourable to health, morals, and religion. This would be accomplished by getting some of them employment in the City, which must necessarily result in better homes and surroundings, or in the opening up for others of a straight course from the Slums to the Farm Colony.

SECTION 2.—THE TRAVELLING HOSPITAL.

Of course, there is only one real remedy for this state of things, and that is to take the people away from the wretched hovels in which they sicken, suffer, and die, with less comfort and consideration than the cattle in the stalls and styes of many a country Squire. And this is certainly our ultimate ambition, but for the present distress something might be done on the lines of district nursing, which is only in very imperfect operation.

I have been thinking that if a little Van, drawn by a pony, could be fitted up with what is ordinarily required by the sick and dying, and trot round amongst these abodes of desolation, with a couple of nurses trained for the business, it might be of immense service, without being very costly. They could have a few simple instruments, so as to draw a tooth or lance an abscess, and what was absolutely requisite for simple surgical operations. A little oil-stove for hot water to prepare a poultice, or a hot foment, or a soap wash, and a number of other necessaries for nursing, could be carried with ease.

The need for this will only be appreciated by those who know how utterly bereft of all the comforts and conveniences for attending to the smallest matters in sickness which prevails in these abodes of wretchedness. It may be suggested, Why don't the people when they are ill go to the hospital? To which we simply reply that they won't. They cling to their own bits of rooms and to the companionship of the members of their own families, brutal as they often are, and would rather stay and suffer, and die in the midst of all the filth and squalor that surrounds them in their own dens, than go to the big house, which, to them, looks very like a prison.

The sufferings of the wretched occupants of the Slums that we have been describing, when sick and unable to help themselves, makes the organisation of some system of nursing them in their own homes a

Christian duty.   Here are a handful of cases, gleaned almost at random from the reports of our Slum Sisters, which will show the value of the agency above described :—

Many of those who are sick have often only one room, and often several children. The Officers come across many cases where, with no one to look after them, they have to lie for hours without food or nourishment of any kind. Sometimes the neighbours will take them in a cup of tea. It is really a mystery how they live.

A poor woman in Drury Lane was paralyzed. She had no one to attend to her ; she lay on the floor, on a stuffed sack, and an old piece of cloth to cover her. Although it was winter, she very seldom had any fire. She had no garments to wear, and but very little to eat.

Another poor woman, who was very ill, was allowed a little money by her daughter to pay her rent and get her food ; but very frequently she had not the strength to light a fire or to get herself food. She was parted from her husband because of his cruelty. Often she lay for hours without a soul to visit or help her.

Adjutant McClellan found a man lying on a straw mattress in a very bad condition. The room was filthy ; the smell made the Officer feel ill. The man had been lying for days without having anything done for him. A cup of water was by his side. The Officers vomited from the terrible smells of this place.

Frequently sick people are found who need the continual application of hot poultices, but who are left with a cold one for hours.

In Marylebone the Officers visited a poor old woman who was very ill. She lived in an underground back kitchen, with hardly a ray of light and never a ray of sunshine. Her bed was made up on some egg boxes. She had no one to look after her, except a drunken daughter, who very often, when drunk, used to knock the poor old woman about very badly. The Officers frequently found that she had not eaten any food up to twelve o'clock, not even a cup of tea to drink. The only furniture in the room was a small table, an old fender, and a box. The vermin seemed to be innumerable.

A poor woman was taken very ill, but, having a small family, she felt she must get up and wash them. While she was washing the baby she fell down and was unable to move. Fortunately a neighbour came in soon after to ask some question, and saw her lying there. She at once ran and fetched another neighbour. Thinking the poor woman was dead, they got her into bed and sent for a doctor. He said she was in consumption and required quiet and nourishment. This the poor woman could not get, on account of her children. She got up a few hours afterwards. As she was going downstairs she fell down again. The neighbour picked her up and put her back to bed, where for a long time she lay thoroughly prostrated. The Officers took her case in hand, fed, and nursed her, cleaned her room and generally looked after her.

In another dark slum the Officers found a poor old woman in an underground back kitchen. She was suffering with some complaint. When they knocked at the door she was terrified for fear it was the landlord. The room was in a most filthy condition, never having been cleaned. She had a penny paraffin lamp which filled the room with smoke. The old woman was at times totally unable to do anything for herself. The Officers looked after her.

Section 3.—REGENERATION OF OUR CRIMINALS.—THE PRISON GATE BRIGADE

Our Prisons ought to be reforming institutions, which should turn men out better than when they entered their doors. As a matter of fact they are often quite the reverse. There are few persons in this world more to be pitied than the poor fellow who has served his first term of imprisonment or finds himself outside the gaol doors without a character, and often without a friend in the world. Here, again, the process of centralization, gone on apace of late years, however desirable it may be in the interests of administration, tells with disastrous effects on the poor wretches who are its victims.

In the old times, when a man was sent to prison, the gaol stood within a stone's throw of his home. When he came out he was at any rate close to his old friends and relations, who would take him in and give him a helping hand to start once more a new life. But what has happened owing to the desire of the Government to do away with as many local gaols as possible? The prisoners, when convicted, are sent long distances by rail to the central prisons, and on coming out find themselves cursed with the brand of the gaol bird, so far from home, character gone, and with no one to fall back upon for counsel, or to give them a helping hand. No wonder it is reported that vagrancy has much increased in some large towns on account of discharged prisoners taking to begging, having no other resource.

In the competition for work no employer is likely to take a man who is fresh from gaol; nor are mistresses likely to engage a servant whose last character was her discharge from one of Her Majesty's prisons. It is incredible how much mischief is often done by well-meaning persons, who, in struggling towards the attainment of an excellent end—such, for instance, as that of economy and efficiency in prison administration—forget entirely the bearing which their reforms may have upon the prisoners themselves.

The Salvation Army has at least one great qualification for dealing with this question. I believe I am in the proud position of being at the head of the only religious body which has always some of its members in gaol for conscience' sake. We are also one of the few religious bodies which can boast that many of those who are in our ranks have gone through terms of penal servitude. We, therefore, know the prison at both ends. Some men go to gaol because they are better than their neighbours, most men because they are worse. Martyrs, patriots, reformers of all kinds belong to the first category. No great cause has ever achieved a triumph before it has furnished a certain quota to the prison population. The repeal of an unjust law is seldom carried until a certain number of those who are labouring for the reform have experienced in their own persons the hardships of fine and imprisonment. Christianity itself would never have triumphed over the Paganism of ancient Rome had the early Christians not been enabled to testify from the dungeon and the arena as to the sincerity and serenity of soul with which they could confront their persecutors, and from that time down to the successful struggles of our people for the right of public meeting at Whitchurch and elsewhere, the Christian religion and the liberties of men have never failed to demand their quota of martyrs for the faith.

When a man has been to prison in the best of causes he learns to look at the question of prison discipline with a much more sympathetic eye for those who are sent there, even for the worst offences, than judges and legislators who only look at the prison from the outside. "A fellow-feeling makes one wondrous kind," and it is an immense advantage to us in dealing with the criminal classes that many of our best Officers have themselves been in a prison cell. Our people, thank God, have never learnt to regard a prisoner as a mere convict—A 234. He is ever a human being to them, who is to be cared for and looked after as a mother looks after her ailing child. At present there seems to be but little likelihood of any real reform in the interior of our prisons. We have therefore to wait until the men come outside, in order to see what can be done. Our work begins when that of the prison authorities ceases. We have already had a good deal of experience in this work, both here and in Bombay, in Ceylon, in South Africa, in Australia and elsewhere, and as the nett result of our experience we proceed now to set forth the measures we intend to adopt, some of which are already in successful operation.

1. We propose the opening of Homes for this class as near as possible to the different gaols.   One for men has just been taken at King's Cross, and will be occupied as soon as it can be got ready. One for women must follow immediately.   Others will be required in different parts of the Metropolis, and contiguous to each of its great prisons.   Connected with these Homes will be workshops in which the inmates will be regularly employed until such time as we can get them work elsewhere.   For this class must also work, not only as a discipline, but as the means for their own support.

2. In order to save, as far as possible, first offenders from the contamination of prison life, and to prevent the formation of further evil companionships, and the recklessness which follows the loss of character entailed by imprisonment, we would offer, in the Police and Criminal Courts, to take such offenders under our wing as were anxious to come and willing to accept our regulations.   The confidence of both magistrates and prisoners would, we think, soon be secured, the friends of the latter would be mostly on our side, and the probability, therefore, is that we should soon have a large number of cases placed under our care on what is known as "suspended sentence," to be brought up for judgment when called upon, the record of each sentence to be wiped out on report being favourable of their conduct in the Salvation Army Home.

3. We should seek access to the prisons in order to gain such acquaintance with the prisoners as would enable us the more effectually to benefit them on their discharge.   This privilege, we think, would be accorded us by the prison authorities when they became acquainted with the nature of our work and the remarkable results which followed it.   The right of entry into the gaols has already been conceded to our people in Australia, where they have free access to, and communion with, the inmates while undergoing their sentences.   Prisoners are recommended-to come to us by the gaol authorities, who also forward to our people information of the date and hour when they leave, in order that they may be met on their release.

4. We propose to meet the criminals at the prison gates with the offer of immediate admission to our Homes.   The general rule is for them to be met by their friends or old associates, who ordinarily belong to the same class.   Any way, it would be an exception to the rule were they not all alike believers in the comforting and cheering power of the intoxicating cup.   Hence the public-house is invariably

adjourned to, where plans for further crime are often decided upon straight away, resulting frequently, before many weeks are past, in the return of the liberated convict to the confinement from which he has just escaped. Having been accustomed during confinement to the implicit submission of themselves to the will of another, the newly-discharged prisoner is easily influenced by whoever first gets hold of him. Now, we propose to be beforehand with these old companions by taking the gaol-bird under our wing and setting before him an open door of hope the moment he crosses the threshold of the prison, assuring him that if he is willing to work and comply with our discipline, he never need know want any more.

5. We shall seek from the authorities the privilege of supervising and reporting upon those who are discharged with tickets-of-leave, so as to free them from the humiliating and harassing duty of having to report themselves at the police stations.

6. We shall find suitable employment for each individual. If not in possession of some useful trade or calling we will teach him one.

7. After a certain length of residence in these Homes, if consistent evidence is given of a sincere purpose to live an honest life, he will be transferred to the Farm Colony, unless in the meanwhile friends or old employers take him off our hands, or some other form of occupation is obtained, in which case he will still be the object of watchful care.

We shall offer to all the ultimate possibility of being restored to Society in this country, or transferred to commence life afresh in another.

With respect to results we can speak very positively, for although our operations up to the present, except for a short time some three years ago, have been limited, and unassisted by the important accessories above described, yet the success that has attended them has been most remarkable. The following are a few instances which might be multiplied :—

J. W. was met at prison gate by the Captain of the Home and offered help. He declined to come at once as he had friends in Scotland who he thought would help him ; but if they failed, he promised to come. It was his first conviction, and he had six months for robbing his employer. His trade was that of a baker. In a few days he presented himself at the Home, and was received. In the course of a few weeks, he professed conversion, and gave every evidence of the change. For four months he was cook and baker in the kitchen, and at last a situation as second hand was offered for him, with the

J. S. Sergeant-major of the Congress Hall Corps. That is three years ago. He is there to-day, saved, and satisfactory ; a thoroughly useful and respectable man.

J. P. was an old offender. He was met at Millbank on the expiration of his last term (five years), and brought to the Home, where he worked at his trade— a tailor. Eventually he got a situation, and has since married. He has now a good home, the confidence of his neighbours, is well saved, and a soldier of the Hackney Corps.

C. M. Old offender, and penal servitude case. Was induced to come to the Home, got saved, was there for a long period, offered for the work, and went into the Field, was Lieutenant for two years, and eventually married. He is now a respectable mechanic and soldier of a Corps in Derbyshire.

J. W. Was manager in a large West End millinery establishment. He was sent out with two ten-pound packages of silver to change. On his way he met a companion and was induced to take a drink. In the tavern the companion made an excuse to go outside and did not return, and W. found one of the packages had been abstracted from his outside pocket. He was afraid to return, and decamped with the other into the country. Whilst in a small town he strolled into a Mission Hall ; there happened to be a hitch in the proceedings, the organist was absent, a volunteer was called or, and W., being a good musician, offered to play. It seems the music took hold of him. In the middle of the hymn he walked out and went to the police station and gave himself up. He got six months. When he came out, he saw that Happy George, an ex-gaol bird, was announced at the Congress Hall. He went to the meeting and was induced to come to the Home. He eventually got saved, and to-day he is at the head of a Mission work in the provinces.

" Old Dan " was a penal servitude case, and had had several long sentences. He came into the Home and was saved. He managed the bootmaking there for a long time. He has since gone into business at Hackney, and is married. He is of four years' standing, a thorough respectable tradesman, and a Salvationist.

Charles C. has done in the aggregate twenty-three years' penal servitude. Was out on licence, and got saved at the Hull Barracks. At that time he had neglected to report himself, and had destroyed his licence, taking an assumed name. When he got saved he gave himself up, and was taken before the magistrate, who, instead of sending him back to fulfil his sentence, gave him up to the Army. He was sent to us from Hull by our representative, is now in our factory and doing well. He is still under police supervision for five years.

H. Kelso. Also a licence man. He had neglected to report himself, and was arrested. While before the magistrate he said he was tired of dishonesty, and would go to the Salvation Army if they would discharge him. He was sent

M

back to penal servitude. Application was made by us to the Home Secretary on his behalf, and Mr. Matthews granted his release. He was handed over to our Officers at Bristol, brought to London, and is now in the Factory, saved and doing well.

E. W. belongs to Birmingham, is in his forty-ninth year, and has been in and out of prison all his life. He was at Redhill Reformatory five years, and his last term was five years' penal servitude. The Chaplain at Pentonville advised him if he really meant reformation to seek the Salvation Army on his release. He came to Thames Street, was sent to the Workshop and professed salvation the following Sunday at the Shelter. This is three months ago. He is quite satisfactory, industrious, contented and seemingly godly.

A. B., Gentleman loafer, good prospects, drink and idleness broke up his home, killed his wife, and got him into gaol. Presbyterian minister, friend of his family, tried to reclaim him, but unsuccessfully. He entered the Prison Gate Home, became thoroughly saved, distributed handbills for the Home, and ultimately got work in a large printing and publishing works, where, after three years' service, he now occupies a most responsible position. Is an elder in the Presbyterian Church, restored to his family, and the possessor of a happy home.

W. C., a native of London, a good-for-nothing lad, idle and dissolute. When leaving England his father warned him that if he didn't alter he'd end his days on the gallows. Served various sentences on all sorts of charges. Over six years ago we took him in hand, admitted him into Prison Gate Brigade Home, where he became truly saved; he got a job of painting, which he had learnt in gaol, and has married a woman who had formerly been a procuress, but had passed through our Rescued Sinners' Home, and there became thoroughly converted. Together they have braved the storms of life, both working diligently for their living. They have now a happy little home of their own, and are doing very well.

F. X., the son of a Government officer, a drunkard, gambler, forger, and all-round blackguard; served numerous sentences for forgery. On his last discharge was admitted into Prison Gate Brigade Home, where he stayed about five months and became truly saved. Although his health was completely shattered from the effects of his sinful life, he steadfastly resisted all temptations to drink, and kept true to God. Through advertising in the *War Cry*, he found his lost son and daughter, who are delighted with the wonderful change in their father. They have become regular attendants at our meetings in the Temperance Hall. He now keeps a coffee-stall, is doing well, and properly saved.

G. A., 72, spent 23 years in gaol, last sentence two years for burglary; was a drunkard, gambler, and swearer. Met on his discharge by the Prison Gate Brigade, admitted into Home, where he remained four months, and became truly saved. He is living a consistent, godly life, and is in employment.

C. D., aged 64, opium-smoker, gambler, blackguard, separated from wife and family, and eventually landed in gaol, was met on his discharge and admitted into Prison Gate Brigade Home, was saved, and is now restored to his wife and family, and giving satisfaction in his employment.

S. T. was an idle, loafing, thieving, swearing, disreputable young man, who lived, when out of gaol, with the low prostitutes of Little Bourke Street. Was taken in hand by our Prison Gate Brigade Officers, who got him saved, then found him work. After a few months he expressed a desire to work for God, and although a cripple, and having to use a crutch, such was his earnestness that he was accepted and has done good service as an Army officer. His testimony is good and his life consistent. He is, indeed, a marvel of Divine grace.

M. J., a young man holding a high position in England, got into a fast set; thought a change to the Colonies would be to his advantage. Started for Australia with £200 odd, of which he spent a good portion on board ship in drink, soon dissipated the balance on landing, and woke up one morning to find himself in gaol, with delirium tremens on him, no money, his luggage lost, and without a friend on the whole continent. On his discharge he entered our Prison Gate Home, became converted, and is now occupying a responsible position in a Colonial Bank.

B. C., a man of good birth, education, and position; drank himself out of home and friends and into gaol, on leaving which he came to our Home; was saved, exhibiting by an earnest and truly consistent life the depth of his conversion, being made instrumental while with us in the salvation of many who, like himself, had come to utter destitution and crime through drink. He is now in a first-class situation, getting £300 a year, wife and family restored, the possessor of a happy home, and the love of God shed abroad in it.

I do not produce these samples, which are but a few, taken at random from the many, for the purpose of boasting. The power which has wrought these miracles is not in me nor in my Officers; it is power which comes down from above. But I think I may fairly point to these cases, in which our instrumentality has been blessed, to the plucking of these brands from the burning, as affording some justification for the plea to be enabled to go on with this work on a much more extended scale. If any other organisation, religious or secular, can show similar trophies as the result of such limited operations as ours have hitherto been among the criminal population, I am willing to give place to them. All that I want is to have the work done.

Section 4.—EFFECTUAL DELIVERANCE FOR THE DRUNKARD.

The number, misery, and hopeless condition of the slaves of strong drink, of both sexes, have been already dealt with at considerable length.

We have seen that there are in Great Britain half a million of men and women, or thereabouts, completely under the domination of this cruel appetite. The utter helplessness of Society to deal with the drunkard has been proved again and again, and confessed on all hands by those who have had experience on the subject As we have before said, the general feeling of all those who have tried their hands at this kind of business is one of despair. They think the present race of drunkards must be left to perish, that every species of effort having proved vain, the energies expended in the endeavour to rescue the parents will be laid out to greater advantages upon the children.

There is a great deal of truth in all this. Our own efforts have been successful in a very remarkable degree. Some of the bravest, most devoted, and successful workers in our ranks are men and women who were once the most abject slaves of the intoxicating cup. ~ Instances of this have been given already. We might multiply them by thousands. Still, when compared with the ghastly array which the drunken army presents to-day, those rescued are comparatively few. The great reason for this is the simple fact that the vast majority of those addicted to the cup are its veritable slaves. No amount of reasoning, or earthly or religious considerations, can have any effect upon a man who is so completely under the mastery of this passion that he cannot break away from it, although he sees the most terrible consequences staring him in the face.

The drunkard promises and vows, but promises and vows in vain. Occasionally he will put forth frantic efforts to deliver himself, but only to fall again in the presence of the opportunity. The insatiable crave controls him. He cannot get away from it. It compels him to drink, whether he will or not, and, unless delivered by an Almighty hand, he will drink himself into a drunkard's grave and a drunkard's hell.

Our annals team with successful rescues effected from the ranks of the drunken army. The following will not only be examples of this, but will tend to illustrate the strength and madness of the passion which masters the slave to strong drink.

Barbara.—She had sunk about as low as any woman could when we found her.

From the age of eighteen, when her parents had forced her to throw over her sailor sweetheart and marry a man with "good prospects," she had been going steadily down.

She did not love her husband, and soon sought comfort from the little public-house only a few steps from her own door. Quarrels in her home quickly gave place to fighting, angry curses, and oaths, and soon her life became one of the most wretched in the place. Her husband made no pretence of caring for her, and when she was ill and unable to earn money by selling fish in the streets, he would go off for a few months, leaving her to keep the house and support herself and babies as best she could. Out of her twenty years of married life, ten were spent in these on-and-off separations. And so she got to live for only one thing—drink. It was life to her; and the mad craving grew to be irresistible. The woman who looked after her at the birth of her child refused to fetch her whisky, so when she had done all she could and left the mother to rest, Barbara crept out of bed and crawled slowly down the stairs over the way to the tap-room, where she sat drinking with the baby, not yet an hour old, in her arms. So things went on, until her life got so unbearable that she determined to have done with it. Taking her two eldest children with her, she went down to the bay, and deliberately threw them both into the water, jumping in herself after them. "Oh, mither mither, dinna droon me!" wailed her little three-year-old Sarah, but she was determined and held them under the water, till, seeing a boat put out to the rescue she knew that she was discovered. Too late to do it now, she thought, and, holding both children, swam quickly back to the shore. A made-up story about having fallen into the water satisfied the boatman, and Barbara returned home dripping and baffled. But little Sarah did not recover from the shock, and after a few weeks her short life ended, and she was laid in the Cemetery.

Yet another time, goaded to desperation, she tried to take her life by hanging herself, but a neighbour came in and cut her down unconscious, but still living. She became a terror to all the neighbourhood, and her name was the bye-word for daring and desperate actions. But our Open-Air Meetings attracted her, she came to the Barracks, got saved, and was delivered from her love of drink and sin.

From being a dread her home became a sort of house of refuge in the little low street where she lived; other wives as unhappy as herself would come in for advice and help. Anyone knew that Barbie was changed, and loved to do all she could for her neighbours. A few months ago she came up to the Captain's in great distress over a woman who lived just opposite. She had been cruelly kicked and cursed by her husband, who had finally bolted the door against her, and she had turned to Barbie as the only hope. And of course Barbie took her in, with her rough-and-ready kindness got her to bed, kept out the other women who crowded round to sympathise and declaim against the husband's brutality, was both nurse and doctor for the poor woman till her child was born and laid in the mother's arms. And then, to Barbie's distress, she could do no more, for the woman, not daring to be absent longer, got up as best she could, and crawled on hands and knees down the little steep steps, across the street, and back to her own door. "But, Barbie!" exclaimed the Captain, horrified, "you should have nursed her, and kept her until she was strong enough." But Barbie answered by reminding the Captain of "John's" fearful temper, and how it might cost the woman her life to be absent from her home more than a couple of hours.

The second is the case of—

Maggie.—She had a home, but seldom was sober enough to reach it at nights. She would fall down on the doorsteps until found by some passer-by or a policeman.

In one of her mad freaks a boon-companion happened to offend her. He was a little hunch-back, and a fellow-drunkard; but without a moment's hesitation, Maggie seized him and pushed him head-foremost down the old-fashioned wide sewer of the Scotch town. Had not some one seen his heels kicking out and rescued him, he would surely have been suffocated.

One winter's night Maggie had been drinking heavily, fighting, too, as usual, and she staggered only as far, on her way home, as the narrow chain-pier. Here she stumbled and fell, and lay along on the snow, the blood oozing from her cuts, and her hair spread out in a tangled mass.

At 5 in the morning, some factory girls, crossing the bridge to their work, came upon her, lying stiff and stark amidst the snow and darkness.

To rouse her from her drunken sleep was hard, but to raise her from the ground was still harder. The matted hair and blood had frozen fast to the earth, and Maggie was a prisoner. After trying to free her in different ways, and receiving as a reward volleys of abuse and bad language, one of the girls ran for a kettle of boiling water, and by pouring it all around her, they succeeded by degrees in *melting* her on to her feet again !

But she came to our Barracks, and got soundly converted, and the Captain was rewarded for nights and days of toil by seeing her a saved and sober woman.

All went right till a friend asked her to his house, to drink his health, and that of his newly-married wife.

"I wouldn't ask you to take anything strong," he said. "Drink to me with this lemonade."

And Maggie, nothing suspecting, drank, and as she drank tasted in the glass her old enemy, whisky!

The man laughed at her dismay, but a friend rushed off to tell the Captain.

"I may be in time, she has not really gone back"; and the Captain ran to the house, tying her bonnet strings as she ran.

"It's no good—keep awa'—I don't want to see'er, Captain," wailed Maggie; "let me have some more—oh, I'm on fire inside."

But the Captain was firm, and taking her to her home, she locked herself in with the woman, and sat with the key in her pocket, while Maggie, half mad with craving, paced the floor like a caged animal, threatening and entreating by terms.

"Never while I live," was all the answer she could get; so she turned to the door, and busied herself there a moment or two. A clinking noise. The Captain started up—to see the door open and Maggie rush through it! Accustomed to stealing and all its "dodges," she had taken the lock off the door, and was away to the nearest public-house.

Down the stairs, Captain after her, into the gin palace; but before the astonished publican could give her the drink she was clamouring for, the "bonnet" was by her side, "If you dare to serve her, I'll break the glass before it reaches her lips. She shall not have any!" and so Maggie was coaxed away, and shielded till the passion was over, and she was herself once more.

But the man who gave her the whisky durst not leave his house for weeks. The roughs got to know of the trap he had laid for her, and would have lynched him could they have got hold of him.

The third is the case of Rose.

Rose was ruined, deserted, and left to the streets when only a girl of thirteen, by a once well-to-do man, who is now, we believe, closing his days in a workhouse in the North of England.

Fatherless, motherless, and you might almost say friendless, Rose trod the broad way to destruction, with all its misery and shame, for twelve long years. Her wild, passionate nature, writhing under the wrong suffered, sought forgetfulness in the intoxicating cup, and she soon became a notorious drunkard. Seventy-four times during her career she was dragged before the magistrates, and seventy-four times, with one exception, she was punished, but the seventy-fourth time she was as far off reformation as ever. The one exception happened on the Queen's Jubilee Day. On seeing her well-known face again before him, the magistrate enquired, "How many times has this woman been here before?"

The Police Superintendent answered, "Fifty times." The magistrate remarked, in somewhat grim humour, "Then this is her Jubilee," and, moved by the coincidence, he let her go free. So Rose spent her jubilee out of prison.

It is a wonder that the dreadful, drunken, reckless, dissipated life she lived did not hurry her to an early grave; it did affect her reason, and for three weeks she was locked up in Lancaster Lunatic Asylum, having really gone mad through drink and sin.

In evidence of her reckless nature, it is said that after her second imprisonment she vowed she would never again walk to the police station; consequently, when in her wild orgies the police found it necessary to arrest her, they had to get her to the police station as best they could, sometimes by requisitioning a wheelbarrow or a cart, or the use of a stretcher, and sometimes they had to carry her right out. On one occasion, towards the close of her career, when driven to the last-named method, four policemen were carrying her to the station, and she was extra violent, screaming, plunging and biting, when, either by accident or design, one of the policemen let go of her head, and it came in contact with the curbstone, causing the blood to pour forth in a stream. As soon as they placed her in the cell the poor creature caught the blood in her hands, and literally washed her face with it. On the following morning she presented a pitiable sight, and before taking her into the court the police wanted to wash her, but she declared she would draw any man's blood who attempted to put a finger upon her; they had spilt her blood, and she would carry it into the court as a witness against them. On coming out of gaol for the last time, she met with a few Salvationists beating the drum and singing "Oh! the Lamb, the bleeding Lamb; He was found worthy." Rose, struck with the song, and impressed with the very faces of the people, followed them, saying to herself, "I never before heard anything like that, or seen such happy looking people." She came into the Barracks; her heart was broken; she found her way to the Penitent Form, and Christ, with His own precious blood, washed her sins away. She arose from her knees and said to the Captain, "It is all right now."

Three months after her conversion a great meeting was held in the largest hall in the town, where she was known to almost every inhabitant. There were about three thousand people present. Rose was called upon to give her testimony to the power of God to save. A more enthusiastic wave of sympathy never greeted any speaker than that which met her from that crowd, every one of whom was familiar with her past history. After a few broken words, in which she spoke of the wonderful change that had taken place, a cousin, who, like herself, had lived a notoriously evil life, came to the Cross.

Rose is now *War Cry* sergeant. She goes into the brothels and gin palaces and other haunts of vice, from which she was rescued, and sells more papers than any other Soldier.

The Superintendent of Police, soon after her conversion, told the Captain of the Corps that in rescuing Rose a more wonderful work had been done than he had seen in all the years gone by.

S. was a native of Lancashire, the son of poor, but pious, parents. He was saved when sixteen years of age. He was first an Evangelist, then a City Missionary for five or six years, and afterwards a Baptist Minister. He then fell under the influence of drink, resigned, and became a commercial traveller, but lost his berth through drink. He was then an insurance agent, and rose to be superintendent, but was again dismissed through drink. During his drunken career he had delirium tremens four times, attempted suicide three times, sold up six homes, was in the workhouse with his wife and family three times. His last contrivance for getting drink was to preach mock sermons, and offer mock prayers in the tap-rooms.

After one of these blasphemous performances in a public-house, on the words, " Are you Saved ? " he was challenged to go to the Salvation Barracks. He went, and the Captain, who knew him well, at once made for him, to plead for his soul, but S. knocked him down, and rushed back to the public-house for more drink. He was, however, so moved by what he had heard that he was unable to raise the liquor to his mouth, although he made three attempts. He again returned to the meeting, and again quitted it for the public-house. He could not rest, and for the third time he returned to the Barracks. As he entered the last time the Soldiers were singing :—

> " Depth of mercy, can there be
> Mercy still reserved for me ?
> Can my God his wrath forbear ?
> Me, the chief of Sinners, spare ? "

This song impressed him still further ; he wept, and remained in the Barracks under deep conviction until midnight. He was drunk all the next day, vainly trying to drown his convictions. The Captain visited him at night, but was quickly thrust out of the house. He was there again next morning, and prayed and talked with S. for nearly two hours. Poor S. was in despair. He persisted that there was no mercy for him. After a long struggle, however, hope sprung up, he fell upon his knees, confessed his sins, and obtained forgiveness.

When this happened, his furniture consisted of a soap-box for a table, and starch boxes for chairs. His wife, himself, and three children, had not slept in a bed for three years. He has now a happy family, a comfortable home, and has been the means of leading numbers of other slaves of sin to the Saviour, and to a truly happy life.

Similar cases, describing the deliverance of drunkards from the bondage of strong drink, could be produced indefinitely. There are Officers marching in our ranks to-day, who where once gripped by this fiendish fascination, who have had their fetters broken, and are now free men in the Army.   Still the mighty torrent of Alcohol, fed by ten thousand manufactories, sweeps on, bearing with it, I have no hesitatiun in saying, the foulest, bloodiest tide that ever flowed from earth to eternity.   The Church of the living God ought not—and to say nothing about religion, the people who have any humanity ought not, to rest without doing something desperate to rescue this half of a million who are in the eddying mael-strom.   We purpose, therefore, the taking away of the people from the temptation which they cannot resist.   We would to God that the temptation could be taken away from them, that every house licensed to send forth the black stream of bitter death were closed, and closed for ever.   But this will not be, we fear, for the present at least.

While in one case drunkenness may be resolved into a habit, in another it must be accounted a disease.   What is wanted in the one case, therefore, is some method of removing the man out of the sphere of the temptation, and in the other for treating the passion as a disease, as we should any other physical affection, bringing to bear upon it every agency, hygienic and otherwise, calculated to effect a cure.

The Dalrymple Homes, in which, on the order of a magistrate and by their own consent, Inebriates can be confined for a time, have been a partial success in dealing with this class in both these respects ; but they are admittedly too expensive to be of any service to the poor.   It could never be hoped that working people of them-selves, or with the assistance of their friends, would be able to pay two pounds a week for the privilege of being removed away from the licensed temptations to drink which surround them at every step. Moreover, could they obtain admission they would feel themselves anything but at ease amongst the class who avail themselves of these institutions.   We propose to establish Homes which will contemplate the deliverance, not of ones and twos, but of multi-tudes, and which will be accessible to the poor, or to persons of any class choosing to use them.   This is our national vice, and it demands nothing short of a national remedy—anyway, one of proportions large enough to be counted national.

1. To begin with, there will be City Homes, into which a man can be taken, watched over, kept out of the way of temptation, and if possible delivered from the power of this dreadful habit.

In some cases persons would be taken in who are engaged in business in the City in the day, being accompanied by an attendant to and from the Home.   In this case, of course, adequate remuneration for this extra care would be required.

2. Country Homes, which we shall conduct on the Dalrymple principle ; that is, taking persons for compulsory confinement, they binding themselves by a bond confirmed by a magistrate that they would remain for a certain period.

The general regulations for both establishments would be something as follows :—

(1).   There would be only one class in each establishment.   It it was found that the rich and the poor did not work comfortably together, separate institutions must be provided.

(2).   All would alike have to engage in some remunerative form of employment.   Outdoor work would be preferred, but indoor employment would be arranged for those for whom it was most suitable, and in such weather and at such times of the year when garden work was impracticable.

(3).   A charge of 10s. per week would be made.   This could be remitted when there was no ability to pay it.

The usefulness of such Homes is too evident to need any discussion.   There is one class of unfortunate creatures who must be objects of pity to all who have any knowledge of their existence, and that is, those men and women who are being continually dragged before the magistrates, of whom we are constantly reading in the police reports, whose lives are spent in and out of prison, at an enormous cost to the country, and without any benefit to themselves.

We should then be able to deal with this class.   It would be possible for a magistrate, instead of sentencing the poor wrecks of humanity to the sixty-fourth and one hundred and twentieth term of imprisonment, to send them to this Institution, by simply remanding them to come up for sentence when called for.   How much cheaper such an arrangement would be for the country !

Section 5.—A NEW WAY OF ESCAPE FOR LOST WOMEN.

THE RESCUE HOMES.

Perhaps there is no evil more destructive of the best interests of Society, or confessedly more difficult to deal with remedially, than that which is known as the Social Evil. We have already seen something of the extent to which this terrible scourge has grown, and the alarming manner in which it affects our modern civilisation.

We have already made an attempt at grappling with this evil, having about thirteen Homes in Great Britain, accommodating 307 girls under the charge of 132 Officers, together with seventeen Homes abroad, open for the same purpose. The whole, although a small affair compared with the vastness of the necessity, nevertheless constitutes perhaps the largest and most efficient effort of its character in the world.

It is difficult to estimate the results that have been already realised. By our varied operations, apart from these Homes, probably hundreds, if not thousands, have been delivered from lives of shame and misery. We have no exact return of the number who have gone through the Homes abroad, but in connection with the work in this country, about 3,000 have been rescued, and are living lives of virtue.

This success has not only been gratifying on account of the blessing it has brought these young women, the gladness it has introduced to the homes to which they have been restored, and the benefit it has bestowed upon Society, but because it has assured us that much greater results of the same character may be realised by operations conducted on a larger scale, and under more favourable circumstances.

With this view we propose to remodel and greatly increase the number of our Homes both in London and the provinces, establishing one in every great centre of this infamous traffic.

To make them very largely Receiving Houses, where the girls will be initiated into the system of reformation, tested as to the reality of their desires for deliverance, and started forward on the highway of truth, virtue, and religion.

From these Homes large numbers, as at present, would be restored to their friends and relatives, while some would be detained in training for domestic service, and others passed on to the Farm Colony.

On the Farm they would be engaged in various occupations. In the Factory, at Bookbinding and Weaving ; in the Garden and Glass-houses amongst fruit and flowers ; in the Dairy, making butter ; in all cases going through a course of House-work which will fit them for domestic service.

At every stage the same process of moral and religious training, on which we specially rely, will be carried forward.

There would probably be a considerable amount of inter-marriage amongst the Colonists, and in this way a number of these girls would be absorbed into Society.

A large number would be sent abroad as domestic servants. In Canada, the girls are taken out of the Rescue Homes as servants, with no other reference than is gained by a few weeks' residence there, and are paid as much as £3 a month wages. The scarcity of domestic servants in the Australian Colonies, Western States of America, Africa, and elsewhere is well known. And we have no doubt that on all hands our girls with 12 months' character will be welcomed, the question of outfit and passage-money being easily arranged for by the persons requiring their services advancing the amount, with an understanding that it is to be deducted out of their first earnings.

Then we have the Colony Over-Sea, which will require the service of a large number. Very few families will go out who will not be very glad to take a young woman with them, not as a menial servant, but as a companion and friend.

By this method we should be able to carry out Rescue work on a much larger scale. At present two difficulties very largely block our way. One is the costliness of the work. The expense of rescuing a girl on the present plan cannot be much less than £7 ; that is, if we include the cost of those with whom we fail, and on whom the money is largely thrown away. Seven pounds is certainly not a very large sum for the measure of benefit bestowed upon the girl by bringing her off the streets, and that which is bestowed on Society by removing her from her evil course. Still, when the work runs into thousands of individuals, the amount required becomes considerable. On the plan proposed we calculate that from the date of their reaching the Farm Colony they will earn nearly all that is required for their support.

The next difficulty which hinders our expansion in this department is the want of suitable and permanent situations. Although
we have been marvellously successful so far, having at this hour
probably 1,200 girls in domestic service alone, still the difficulty in this
respect is great. Families are naturally shy at receiving these poor
unfortunates when they can secure the help they need combined with
unblemished character ; and we cannot blame them.

Then, again, it can easily be understood that the monotony of
domestic service in this country is not altogether congenial to the
tastes of many of these girls, who have been accustomed to a life of
excitement and freedom. This can be easily understood. To be
shut up seven days a week with little or no intercourse, either with
friends or with the outside world, beyond that which comes of the
weekly Church service or " night out " with nowhere to go, as many
of them are tied off from the Salvation Army Meetings, becomes
very monotonous, and in hours of depression it is not to be
wondered at if a few break down in their resolutions, and fall back
into their old ways.

On the plan we propose there is something to cheer these girls forward. Life on the farm will be attractive. From there they can go to
a new country and begin the world afresh, with the possibility of being
married and having a little home of their own some day. With such
prospects, we think, they will be much more likely to fight their
way through seasons of darkness and temptation than as at
present.

This plan will also make the task of rescuing the girls much more
agreeable to the Officers engaged in it. They will have this future
to dwell upon as an encouragement to persevere with the girls, and
will be spared one element at least in the regret they experience,
when a girl falls back into old habits, namely, that she earned the
principal part of the money that has been expended upon her.

That girls can be rescued and blessedly saved even now, despite
all their surroundings, we have many remarkable proofs. Of these
take one or two as examples :—

J. W. was brought by our Officers from a neighbourhood which has, by reason
of the atrocities perpetrated in it, obtained an unenviable renown, even among
similar districts of equally bad character.

She was only nineteen. A country girl. She had begun the struggle for
life early as a worker in a large laundry, and at thirteen years of age was led
away by an inhuman brute. The first false step taken, her course on the

downward road was rapid, and growing restless and anxious for more scope than that afforded in a country town, she came up to London.

For some time she lived the life of extravagance and show, known to many of this class for a short time—having plenty of money, fine clothes, and luxurious surroundings—until the terrible disease seized her poor body, and she soon found herself deserted, homeless and friendless, an outcast of Society.

When we found her she was hard and impenitent, difficult to reach even with the hand of love ; but love won, and since that time she has been in two or three situations, a consistent Soldier of an Army corps, and a champion *War Cry* seller.

A TICKET-OF-LEAVE WOMAN.

A. B. was the child of respectable working people—Roman Catholics—but was early left an orphan. She fell in with bad companions, and became addicted to drink, going from bad to worse until drunkenness, robbery, and harlotry brought her to the lowest depths. She passed seven years in prison, and after the last offence was discharged with seven years' police supervision. Failing to report herself, she was brought before the bench.

The magistrate inquired whether she had ever had a chance in a Home of any kind. " She is too old, no one will take her," was the reply, but a Detective present, knowing a little about the Salvation Army, stepped forward and explained to the magistrate that he did not think the Salvation Army refused any who applied. She was formally handed over to us in a deplorable condition, her clothing the scantiest and dirtiest. For over three years she has given evidence of a genuine reformation, during which time she has industriously earned her own living.

A WILD WOMAN.

In visiting a slum in a town in the North of England, our Officers entered a hole, unfit to be called a human habitation—more like the den of some wild animal almost the only furniture of which was a filthy iron bedstead, a wooden box to serve for table and chair, while an old tin did duty as a dustbin.

The inhabitant of this wretched den was a poor woman, who fled into the darkest corner of the place as our Officer entered. This poor wretch was the victim of a brutal man, who never allowed her to venture outside the door, keeping her alive by the scantiest allowance of food. Her only clothing consisted of a sack tied round her body. Her feet were bare, her hair matted and foul, presenting on the whole such an object as one could scarcely imagine living in a civilised country.

She had left a respectable home, forsaken her husband and family, and sunk so low that the man who then claimed her boasted to the Officer that he had bettered her condition by taking her off the streets.

We took the poor creature away, washed and clothed her ; and, changed in heart and life, she is one more added to the number of those who rise up to bless the Salvation Army workers.

Section 6.—A PREVENTIVE HOME FOR UNFALLEN GIRLS WHEN IN DANGER.

There is a story told likely enough to be true about a young girl who applied one evening for admission to some home established for the purpose of rescuing fallen women. The matron naturally inquired whether she had forfeited her virtue ; the girl replied in the negative. She had been kept from that infamy, but she was poor and friendless, and wanted somewhere to lay her head until she could secure work, and obtain a home. The matron must have pitied her, but she could not help her as she did not belong to the class for whose benefit the Institution was intended. The girl pleaded, but the matron could not alter the rule, and dare not break it, they were so pressed to find room for their own poor unfortunates, and she could not receive her. The poor girl left the door reluctantly but returned in a very short time, and said, " I *am* fallen now, will you take me in ? "

I am somewhat slow to credit this incident ; anyway it is true in spirit, and illustrates the fact that while there are homes to which any poor, ruined, degraded harlot can run for shelter, there is only here and there a corner to which a poor friendless, moneyless, homeless, but unfallen girl can fly for shelter from the storm which bids fair to sweep her away whether she will or no into the deadly vortex of ruin which gapes beneath her.

In London and all our large towns there must be a considerable number of poor girls who from various causes are suddenly plunged into this forlorn condition ; a quarrel with the mistress and sudden discharge, a long bout of disease and dismissal penniless from the hospital, a robbery of a purse, having to wait for a situation until the last penny is spent, and many other causes will leave a girl an almost hopeless prey to the linx-eyed villains who are ever watching to take advantage of innocence when in danger. Then, again, what a number there must be in a great city like London who are ever faced with the alternative of being turned out of doors if they refuse to submit themselves to the infamous overtures of those around them. I understand that the Society for the Protection of Children prosecuted last year a fabulous number of fathers for unnatural sins with their children. If so many were brought to justice, how many were there of whom the world never heard in any shape or form? We have only to imagine how many a poor girl is faced with the terrible alternative of being driven literally into the streets by employers or relatives or others in whose power she is unfortunately placed.

Now, we want a real home for such— a house to which any girl can fly at any hour of the day or night, and be taken in, cared for, shielded from the enemy, and helped into circumstances of safety.

The Refuge we propose will be very much on the same principle as the Homes for the Destitute already described. We should accept any girls, say from fourteen years of age, who were without visible means of support, but who were willing to work, and to conform to discipline. There would be various forms of labour provided, such as laundry work, sewing, knitting by machines, &c. Every beneficial influence within our power would be brought to bear on the rectification and formation of character. Continued efforts would be made to secure situations according to the adaptation of the girls, to restore wanderers to their homes, and otherwise provide for all. From this, as with the other Homes, there will be a way made to the Farm and to the Colony over the sea. The institutions would be multiplied as we had means and found them to be necessary, and made self-supporting as far as possible.

Section 7.—ENQUIRY OFFICE FOR LOST PEOPLE.

Perhaps nothing more vividly suggests the varied forms of broken-hearted misery in the great City than the statement that 18,000 people are lost in it every year, of whom 9,000 are never heard of any more, anyway in this world. What is true about London is, we suppose, true in about the same proportion of the rest of the country. Husbands, sons, daughters, and mothers are continually disappearing, and leaving no trace behind.

In such cases, where the relations are of some importance in the world, they may interest the police authorities sufficiently to make some enquiries in this country, which, however, are not often successful ; or where they can afford to spend large sums of money, they can fall back upon the private detective, who will continue these enquiries, not only at home but abroad.

But where the relations of the missing individual are in humble circumstances, they are absolutely powerless, in nine cases out of ten, to effectually prosecute any search at all that is likely to be successful.

Take, for instance, a cottager in a village, whose daughter leaves for service in a big town or city. Shortly afterwards a letter arrives informing her parents of the satisfactory character of her place. The mistress is kind, the work easy, and she likes her fellow servants. She is going to chapel or church, and the family are pleased. Letters continue to arrive of the same purport, but, at length, they suddenly cease. Full of concern, the mother writes to know the reason, but no answer comes back, and after a time the letters are returned with " gone, no address," written on the envelope. The mother writes to the mistress, or the father journeys to the city, but no further information can be obtained beyond the fact that " the girl has conducted herself somewhat mysteriously of late ; had ceased to be as careful at her work ; had been noticed to be keeping company with some young man ; had given notice and disappeared altogether."

Now, what can these poor people do ? They apply to the police, but they can do nothing. Perhaps they ask the clergyman of the parish, who is equally helpless, and there is nothing for them but for the father to hang his head and the mother to cry herself to sleep—to long, and wait, and pray for information that perhaps never comes, and to fear the worst.

Now, our Enquiry Department supplies a remedy for this state of things. In such a case application would simply have to be made to the nearest Salvation Army Officer—probably in her own village, any way, in the nearest town—who would instruct the parents to write to the Chief Office in London, sending portraits and all particulars. Enquiries would at once be set on foot, which would very possibly end in the restoration of the girl.

The achievements of this Department, which has only been in operation for a short time, and that on a limited scale, as a branch of Rescue Work, have been marvellous. No more romantic stories can be found in the pages of our most imaginative writers than those it records. We give three or four illustrative cases of recent date.

ENQUIRY.

RESULT.

A LOST HUSBAND.

Mrs. S., of New Town, Leeds, wrote to say that ROBERT R. left England in July 1889, for Canada to improve his position. He left a wife and four little children behind, and on leaving said that if he were successful out there he should send for them, but if not he should return.

As he was unsuccessful, he left Montreal in the Dominion Liner "Oregon," on October 30th, but except receiving a card from him ere he started, the wife and friends had heard no more of him from that day till the date they wrote us.

They had written to the "Dominion" Company, who replied that " he landed at Liverpool all right," so, thinking he had disappeared upon his arrival, they put the matter in the hands of the Liverpool Police, who, after having the case in hand for several weeks made the usual report—" Cannot be traced."

We at once commenced looking for some passenger who had come over by the same steamer, and after the lapse of a little time we succeeded in getting hold of one.

In our first interview with him we learned that Robert R. did not land at Liverpool, but when suffering from depression threw himself overboard three days after leaving America, and was drowned. We further elicited that upon his death the sailors rifled his clothes and boxes, and partitioned them.

We wrote the Company reporting this, and they promised to make enquiries and amends, but as too often happens, upon making report of the same to the family they took the matter into their own hands, dealt with the Company direct, and in all probability thereby lost a good sum in compensation which we should probably have obtained for them.

A LOST WIFE.

F. J. L. asked us to seek for his wife, who left him on November 4th, 1888. He feared she had gone to live an immoral life; gave us two addresses at which she might possibly be heard of, and a description. They had three children.

Enquiries at the addresses given elicited no information, but from observation in the neighbourhood the woman's whereabouts was discovered.

After some difficulty our Officer obtained an interview with the woman, who was greatly astonished at our having discovered her. She was dealt with faithfully and firmly: the plain truth of God set before her, and was covered with shame and remorse, and promised to return.

We communicated with Mr. L. A few days after he wrote that he had been telegraphed for, had forgiven his wife, and that they were re-united.

Soon afterwards she wrote expressing her deep gratitude to Mrs. Bramwell Booth for the trouble taken in her case.

A LOST CHILD.

ALICE P. was stolen away from home by Gypsies ten years ago, and now longs to find her parents to be restored to them. She believes her home to be in Yorkshire.

The Police had this case in hand for some time, but failed entirely.

With these particulars we advertised in the "War Cry." Captain Green, seeing the advertisement, wrote, April 3rd, from 3, C. S., M. H., that her Lieutenant knew a family of the name advertised for, living at Gomersal, Leeds.

We, on the 4th, wrote to this address for confirmation.

April 6th, we heard from Mr. P——, that this lass is his child, and he writes full of gratitude and joy, saying he will send money for her to go home. We, meanwhile, get from the Police, who had long sought this girl, a full description and photo, which we sent to Captain Cutmore; and on April 9th, she wrote us to the effect that the girl exactly answered the description. We got from the parents 15/- for the fare, and Alice was once more restored to her parents.

Praise God.

A LOST DAUGHTER.

E. W. Age 17. Application from this girl's mother and brother, who had lost all trace of her since July, 1885, when she left for Canada. Letters had been once or twice received, dated from Montreal, but they stopped.

A photo., full description, and handwriting were supplied.

We discovered that some kind Church people here had helped E. W. to emigrate, but they had no information as to her movements after landing.

Full particulars, with photo., were sent to our Officers in Canada. The girl was not found in Montreal. The information was then sent to Officers in other towns in that part of the Colony.

The enquiry was continued through some months ; and, finally, through our Major of Division, the girl was reported to us as having been recognised in one of our Barracks and identified. When suddenly called by *her own name*, she nearly fainted with agitation.

She was in a condition of terrible poverty and shame, but at once consented, on hearing of her mother's enquiries, to go into one of our Canadian Rescue Homes. She is now doing well.

Her mother's joy may be imagined.

A LOST SERVANT.

Mrs. M., Clevedon, one of Harriett P.'s old mistresses, wrote us, in deep concern, about this girl. She said she was a good servant, but was ruined by the young man who courted her, and had since had three children. Occasionally, she would have a few bright and happy weeks, but would again lapse into the " vile path."

Mrs. M. tells us that Harriett had good parents, who are dead, but she still has a respectable brother in Hampshire. The last she heard of her was that some weeks ago she was staying at a Girl's Shelter at Bristol, but had since left, and nothing more had been heard of her.

The enquirer requested us to find her, and in much faith added, " I believe you are the only people who, if successful in tracing her, can rescue and do her a permanent good."

We at once set enquiries on foot, and in the space of a few days found that she had started from Bristol on the road for Bath. Following her up we found that at a little place called Bridlington, on the way to Bath, she had met a man, of whom she enquired her way. He hearing a bit of her story, after taking her to a public-house, prevailed upon her to go home and live with him, as he had lost his wife.

It was at this stage that we came upon the scene, and having dealt with them both upon the matter, got her to consent to come away if the man would not marry her, giving him two days to make up his mind.

The two days' respite having expired and, he being unwilling to undertake matrimony, we brought her away, and sent her to one of our Homes, where she is enjoying peace and penitence.

When we informed the mistress and brother of the success, they were greatly rejoiced and overwhelmed us with thanks.

A LOST HUSBAND.

In a seaside home last Christmas there was a sorrowing wife, who mourned over the basest desertion of her husband. Wandering from place to place drinking, he had left her to struggle alone with four little ones dependent upon her exertions.

Knowing her distress, the captain of the corps wrote begging us to advertise for the man in the *Cry*. We did this, but for some time heard nothing of the result.

Several weeks later a Salvationist entered a beer-house, where a group of men were drinking, and began to distribute *War Crys* amongst them, speaking here and there upon the eternity which faced everyone.

At the counter stood a man with a pint pot in hand, who took one of the papers passed to him, and glancing carelessly down its columns caught sight of his own name, and was so startled that the pot fell from his grasp to the floor. "Come home," the paragraph ran, "and all will be forgiven."

His sin faced him; the thought of a broken-hearted wife and starving children conquered him completely, and there and then he left the public-house, and started to walk home—a distance of many miles—arriving there about midnight the same night, after an absence of eleven months.

The letter from his wife telling the good news of his return, spoke also of his determination by God's help to be a different man, and they are both attendants at the Salvation Army barracks.

A SEDUCER COMPELLED TO PAY.

Amongst the letters that came to the Inquiry Office one morning was one from a girl who asked us to help her to trace the father of her child who had for some time ceased to pay anything towards its support. The case had been brought into the Police Court, and judgment given in her favour, but the guilty one had hidden, and his father refused to reveal his whereabouts.

We called upon the elder man and laid the matter before him, but failed to prevail upon him either to pay his son's liabilities or to put us into communication with him. The answers to an advertisement in the *War Cry*, however, had brought the required information as to his son's whereabouts, and the same morning that our Inquiry Officer communicated with the police, and served a summons for the overdue money, the young man had also received a letter from his father advising him to leave the country at once. He had given notice to his employers; and the £16 salary he received, with some help his father had sent him towards the journey, he was compelled to hand over to the mother of his child.

FOUND IN THE BUSH.

A year or two ago a respectable-looking Dutch girl might have been seen making her way quickly and stealthily across a stretch of long rank grass towards the shelter of some woods on the banks of a distant river. Behind her lay the South African town from which she had come, betrayed, disgraced, ejected from her home with words of bitter scorn, having no longer a friend in the wide world who would hold out to her a hand of help. What could there be better for her than to plunge into that river yonder, and end this life—no matter what should come after the plunge ? But Greetah feared the "future," and turned aside to spend the night in darkness, wretched and alone.

     *   *   *   *   *

Seven years had passed. An English traveller making his way through Southern Africa halted for the Sabbath at a little village on his route. A ramble through the woods brought him unexpectedly in front of a kraal, at the door of which squatted an old Hottentot, with a fair white-faced child playing on the ground near by. Glad to accept the proffered shelter of the hut from the burning sun, the traveller entered, and was greatly astonished to find within a young white girl, evidently the mother of the frolicsome child. Full of pity for the strange pair, and especially for the girl, who wore an air of refinement little to be expected in this out-of-the-world spot, he sat down on the earthen floor, and told them of the wonderful Salvation of God. This was Greetah, and the Englishman would have given a great deal if he could have rescued her from this miserable lot. But this was impossible, and with reluctance he bid her farewell.

     *   *   *   *   *

It was an English home. By a glowing fire one night a man sat alone, and in his imaginings there came up the vision of the girl he had met in the Hottentot's Kraal, and wondering whether any way of rescue was possible. Then he remembered reading, since his return, the following paragraph in the *War Cry :*—

"TO THE DISTRESSED.

" The Salvation Army invite parents, relations, and friends in any part of the world interested in any woman or girl who is known, or feared to be, living in immorality, or is in danger of coming under the control of immoral persons, to write, stating full particulars, with names, dates, and address of all concerned, and, if possible, a photograph of the person in whom the interest is taken.

" All letters, whether from these persons or from *such women or girls them-selves,* will be regarded as strictly confidential. They may be written in any language, and should be addressed to Mrs. Bramwell Booth, 101, Queen Victoria Street, London, E.C."

" It will do no harm to try, anyhow," exclaimed he, " the thing haunts me as it is," and without further delay he penned an account of his African adventure, as full as possible. The next African mail carried instructions to the Officer in Command of our South African work.

*   *   *   *   *

Shortly after, one of our Salvation Riders was exploring the bush, and after some difficulty the kraal was discovered—the girl was rescued and saved. The Hottentot was converted afterwards, and both are now Salvation Soldiers.

Apart from the independent agencies employed to prosecute this class of enquiries, which it is proposed to very largely increase, the Army possesses in itself peculiar advantages for this kind of investigation. The mode of operation is as follows :—

There is a Head Centre under the direction of a capable Officer and assistants, to which particulars of lost husbands, sons, daughters, and wives, as the case may be, are forwarded. These are advertised, except when deemed inadvisable, in the English " War Cry," with its 300,000 circulation, and from it copied into the twenty-three other " War Crys " published in different parts of the world. Specially prepared information in each case is sent to the local Officers of the Army when that is thought wise, or Special Enquiry Officers trained to their work are immediately set to work to follow up any clue which has been given by enquiring relations or friends.

Every one of its 10,000 Officers, nay, almost every soldier in its ranks, scattered, as they are, through every quarter of the globe, may be regarded as an Agent.

A small charge for enquiries is made, and, where persons are able, all the costs of the investigation will be defrayed by them.

Section 8.—REFUGES FOR THE CHILDREN OF THE STREETS.

For the waifs and strays of the streets of London much commiseration is expressed, and far more pity is deserved than is bestowed. We have no direct purpose of entering on a crusade on their behalf, apart from our attempt at changing the hearts and lives and improving the circumstances of their parents.

Our main hope for these wild, youthful, outcasts lies in this direction. If we can reach and benefit their guardians, morally and materially, we shall take the most effectual road to benefit the children themselves.

Still, a number of them will unavoidably be forced upon us ; and we shall be quite prepared to accept the responsibility of dealing with them, calculating that our organisation will enable us to do so, not only with facility and efficiency, but with trifling cost to the public.

To begin with, Children's Crèches or Children's Day Homes would be established in the centres of every poor population, where for a small charge babies and young children can be taken care of in the day while the mothers are at work, instead of being left to the dangers of the thoroughfares or the almost greater peril of being burnt to death in their own miserable homes.

By this plan we shall not only be able to benefit the poor children, if in no other direction than that of soap and water and a little wholesome food, but exercise some humanising influence upon the mothers themselves.

On the Farm Colony, we should be able to deal with the infants from the Unions and other quarters. Our Cottage mothers, with two or three children of their own, would readily take in an extra one on the usual terms of boarding out children, and nothing would be more simple or easy for us than to set apart some trustworthy experienced dame to make a constant inspection as to whether the children placed out were enjoying the necessary conditions of health and general well-being. Here would be a Baby Farm carried on with the most favourable surroundings.

Section 9.—INDUSTRIAL SCHOOLS.

I also propose, at the earliest opportunity, to give the subject of the industrial training of boys a fair trial; and, if successful, follow it on with a similar one for girls. I am nearly satisfied in my own mind that the children of the streets taken, say at eight years of age, and kept till, say twenty-one, would, by judicious management and the utilisation of their strength and capacity, amply supply all their own wants, and would, I think, be likely to turn out thoroughly good and capable members of the community.

Apart from the mere benevolent aspect of the question, the present system of teaching is, to my mind, unnatural, and shamefully wasteful of the energies of the children. Fully one-half the time that boys and girls are compelled to sit in school is spent to little or no purpose—nay, it is worse than wasted. The minds of the children are only capable of useful application for so many consecutive minutes, and hence the rational method must be to apportion the time of the children ; say, half the morning's work to be given to their books, and the other half to some industrial employment ; the garden would be most natural and healthy in fair weather, while the workshop should be fallen back upon when unfavourable.

By this method health would be promoted, school would be loved, the cost of education would be cheapened, and the natural bent of the child's capacities would be discovered and could be cultivated. Instead of coming out of school, or going away from apprenticeship, with the most precious part of life for ever gone so far as learning is concerned, chained to some pursuit for which there is no predilection, and which promises nothing higher than mediocrity if not failure—the work for which the mind was peculiarly adapted and for which, therefore, it would have a natural capacity, would not only have been discovered, but the bent of the inclination cultivated, and the life's work chosen accordingly.

It is not for me to attempt any reform of our School system on this model. But I do think that I may be allowed to test the theory by its practical working in an Industrial School in connection with the Farm Colony. I should begin probably with children selected for their goodness and capacity, with a view to imparting a superior education, thus fitting them for the position of Officers in all parts of the world, with the special object of raising up a body of men thoroughly trained and educated, among other things, to carry out all the branches of the Social work that are set forth in this book, and it may be to instruct other nations in the same.

Section 10.—ASYLUMS FOR MORAL LUNATICS.

There will remain, after all has been said and done, one problem that has yet to be faced. You may minimise the difficulty every way, and it is your duty to do so, but no amount of hopefulness can make us blink the fact that when all has been done and every chance has been offered, when you have forgiven your brother not only seven times but seventy times seven, when you have fished him up from the mire and put him on firm ground only to see him relapse and again relapse until you have no strength left to pull him out once more, there will still remain a residuum of men and women who have, whether from heredity or custom, or hopeless demoralisation, become reprobates. After a certain time, some men of science hold that persistence in habits tends to convert a man from a being with freedom of action and will into a mere automaton. There are some cases within our knowledge which seem to confirm the somewhat dreadful verdict by which a man appears to be a lost soul on this side of the grave.

There are men so incorrigibly lazy that no inducement that you can offer will tempt them to work; so eaten up by vice that virtue is abhorrent to them, and so inveterately dishonest that theft is to them a master passion. When a human being has reached that stage, there is only one course that can be rationally pursued. Sorrowfully, but remorselessly, it must be recognised that he has become lunatic, morally demented, incapable of self-government, and that upon him, therefore, must be passed the sentence of permanent seclusion from a world in which he is not fit to be at large. The ultimate destiny of these poor wretches should be a penal settlement where they could be confined during Her Majesty's pleasure as are the criminal lunatics at Broadmoor. It is a crime against the race to allow those who are so inveterately depraved the freedom to wander abroad, infect their fellows, prey upon Society, and to multiply their kind. Whatever else Society may do, and suffer to be done, this thing it ought not to allow, any more than it should allow the free perambulation of a mad dog. But before we come to this I would have every possible means tried to effect their reclamation. Let Justice punish them, and Mercy put her arms around them ; let them be appealed to by penalty and by reason, and by every influence, human and Divine, that can possibly be brought to bear upon them. Then, if all alike failed, their ability to further curse their fellows and themselves should be stayed.

They will still remain objects worthy of infinite compassion. They should lead as human a life as is possible to those who have fallen under so terrible a judgment. They should have their own little cottages in their own little gardens, under the blue sky, and, if possible, amid the green fields. I would deny them none of the advantages, moral, mental, and religious which might minister to their diseased minds, and tend to restore them to a better state. Not until the breath leaves their bodies should we cease to labour and wrestle for their salvation But when they have reached a certain point access to their fellow men should be forbidden. Between them and the wide world there should be reared an impassable barrier, which once passed should be recrossed no more for ever. Such a course must be wiser than allowing them to go in and out among their fellows, carrying with them the contagion of moral leprosy, and multiplying a progeny doomed before its birth to inherit the vices and diseased cravings of their unhappy parents.

To these proposals three leading objections will probably be raised

      1. It may be said that to shut out men and women from that liberty which is their universal birthright would be cruel.

To this it might be sufficient to reply that this is already done ; twenty years' immurement is a very common sentence passed upon wrong-doers, and in some cases the law goes as far as to inflict penal servitude for life. But we say further that it would be far more merciful treatment than that which is dealt out to them at present, and it would be far more likely to secure a pleasant existence. Knowing their fate they would soon become resigned to it. Habits of industry, sobriety, and kindness with them would create a restfulness of spirit which goes far on in the direction of happiness, and if religion were added it would make that happiness complete. There might be set continually before them a large measure of freedom and more frequent intercourse with the world in the shape of correspondence, newspapers, and even occasional interviews with relatives, as rewards for well-doing. And in sickness and old age their latter days might be closed in comfort. In fact, so far as this class of people were concerned, we can see that they would be far better circumstanced for happiness in this life and in the life to come than in their present liberty—if a life spent alternatively in drunkenness, debauchery, and crime, on the one hand, or the ⌐rison on the other, can be called liberty.

2. It may be said that the carrying out of such a suggestion would be too expensive.

To this we reply that it would have to be very costly to exceed the expense in which all such characters involve the nation under the present regulations of vice and crime. But there is no need for any great expense, seeing that after the first outlay the inmates of such an institution, if it were fixed upon the land, would readily earn all that would be required for their support.

3. But it may be said that this is impossible.

It would certainly be impossible other than as a State regulation. But it would surely be a very simple matter to enact a law which should decree that after an individual had suffered a certain number of convictions for crime, drunkenness, or vagrancy, he should forfeit his freedom to roam abroad and curse his fellows. When I include vagrancy in this list, I do it on the supposition that the opportunity and ability for work are present. Otherwise it seems to me most heartless to punish a hungry man who begs for food because he can in no other way obtain it. But with the opportunity and ability for work I would count the solicitation of charity a crime, and punish it as such. Anyway, if a man would not work of his own free will I would compel him.

CHAPTER VI.

ASSISTANCE IN GENERAL.

There are many who are not lost, who need help. A little assistance given to-day will perhaps prevent the need of having to save them to-morrow. There are some, who, after they have been rescued, will still need a friendly hand. The very service which we have rendered them at starting makes it obligatory upon us to finish the good work. Hitherto it may be objected that the Scheme has dealt almost exclusively with those who are more or less disreputable and desperate. This was inevitable. We obey our Divine Master and seek to save those who are lost. But because, as I said at the beginning, urgency is claimed rightly for those who have no helper, we do not, therefore, forget the needs and the aspirations of the decent working people who are poor indeed, but who keep their feet, who have not fallen, and who help themselves and help each other. They constitute the bulk of the nation. There is an uppercrust and a submerged tenth. But the hardworking poor people, who earn a pound a week or less, constitute in every land the majority of the population. We cannot forget them, for we are at home with them. We belong to them and many thousands of them belong to us. We are always studying how to help them, and we think this can be done in many ways, some of which I proceed to describe.

Section i.—IMPROVED LODGINGS.

The necessity for a superior class of lodgings for the poor men rescued at our Shelters has been forcing itself already upon our notice, and demanding attention. One of the first things that happens when a man, lifted out of the gutter, has obtained a situation, and is earning a decent livelihood, is for him to want some better accommodation than that afforded at the Shelters. We have some hundreds on our hands now who can afford to pay for greater comfort and seclusion. These are continually saying to us something like the following :—

"The Shelters are all very well when a man is down in his luck. They have been a good thing for us; in fact, had it not been for them, we would still have been without a friend, sleeping on the Embankment, getting our living dishonestly, or not getting a living at all. We have now got work, and want a bed to sleep on, and a room to ourselves, and a box, or something where we can stow away our bits of things. Cannot you do something for us?" We have replied that there were Lodging-houses elsewhere, which, now that they were in work, they could afford to pay for, where they would obtain the comfort they desired. To this they answer, "That is all very well. We know there are these places, and that we could go to them. But then," they said, "you see, here in the Shelters are our mates, who think as we do. And there is the prayer, and the meeting, and kind influence every night, that helps to keep us straight. We would like a better place, but if you cannot find us one we would rather stop in the Shelter and sleep on the floor, as we have been doing, than go to something more complete, get into bad company, and so fall back again to where we were before."

But this, although natural, is not desirable ; for, if the process went on, in course of time the whole of the Shelter Depôts would be taken up by persons who had risen above the class for whom they were originally destined. I propose, therefore, to draft those who get on, but wish to continue in connection with the Army, into a superior lodging-house, a sort of

POOR MAN'S METROPOLE,

managed on the same principles, but with better accommodation in every way, which, I anticipate, would be self-supporting from the first. In these homes there would be separate dormitories, good sitting-rooms, cooking conveniences, baths, a hall for meetings, and many other comforts, of which all would have the benefit at as low a figure above cost price as will not only pay interest on the original outlay, but secure us against any shrinkage of capital.

Something superior in this direction will also be required for the women. Having begun, we must go on. Hitherto I have proposed to deal only with single men and single women, but one of the consequences of getting hold of these men very soon makes itself felt. Your ragged, hungry, destitute Out-of-Work in almost every case is married. When he comes to us he comes as single and is dealt with as such, but after you rouse in him aspirations for better things he remembers the wife whom he has probably enough deserted, or left from sheer inability to provide her anything to eat. As soon as such a man finds himself under good influence and fairly employed his first thought is to go and look after the "Missis." There is very little reality about any change of heart in a married man who does not thus turn in sympathy and longing towards his wife, and the more successful we are in dealing with these people the more inevitable it is that we shall be confronted with married couples who in turn demand that we should provide for them lodgings. This we propose to do also on a commercial footing. I see greater developments in this direction, one of which will be described in the chapter relating to Suburban Cottages. The Model-lodging House for Married People is, however, one of those things that must be provided as an adjunct of the Food and Shelter Depôts.

SECTION 2.—MODEL SUBURBAN VILLAGES.

As I have repeatedly stated already, but will state once more, for it is important enough to bear endless repetition, one of the first steps which must inevitably be taken in the reformation of this class, is to make for them decent, healthy, pleasant homes, or help them to make them for themselves, which, if possible, is far better. I do not regard the institution of any first, second, or third-class lodging-houses as affording anything but palliatives of the existing distress. To substitute life in a boarding-house for life in the streets is, no doubt, an immense advance, but it is by no means the ultimatum. Life in a boarding-house is better than the worst, but it is far from being the best form of human existence. Hence, the object I constantly keep in view is how to pilot those persons who have been set on their feet again by means of the Food and Shelter Depôts, and who have obtained employment in the City, into the possession of homes of their own.

Neither can I regard the one, or at most two, rooms in which the large majority of the inhabitants of our great cities are compelled to spend their days, as a solution of the question. The over-crowding which fills every separate room of a tenement with a human litter, and compels family life from the cradle to the grave to be lived within the four walls of a single apartment, must go on reproducing in endless succession all the terrible evils which such a state of things must inevitably create.

Neither can I be satisfied with the vast, unsightly piles of barrack-like buildings, which are only a slight advance upon the Union Bastille—dubbed Model Industrial Dwellings—so much in fashion at present, as being a satisfactory settlement of the burning question of the housing of the poor.

As a contribution to this question, I propose the establishment of a series of Industrial Settlements or Suburban Villages, lying out in the country, within a reasonable distance of all our great cities, composed of cottages of suitable size and construction, and with all needful comfort and accommodation for the families of working-men, the rent of which, together with the railway fare, and other economic conveniences, should be within the reach of a family of moderate income.

This proposal lies slightly apart from the scope of this book, otherwise I should be disposed to elaborate the project at greater length. I may say, however, that what I here propose has been carefully thought out, and is of a perfectly practical character. In the planning of it I have received some valuable assistance from a friend who has had considerable experience in the building trade, and he stakes his professional reputation on its feasibility. The following, however, may be taken as a rough outline :—

The Village should not be more than twelve miles from town ; should be in a dry and healthy situation, and on a line of railway. It is not absolutely necessary that it should be near a station, seeing that the company would, for their own interests, immediately erect one.

The Cottages should be built of the best material and workmanship. This would be effected most satisfactorily by securing a contract for the labour only, the projectors of the Scheme purchasing the materials and supplying them direct from the manufacturers to the builders. The cottages would consist of three or four rooms, with a scullery, and out-building in the garden. The cottages should be built in terraces, each having a good garden attached.

Arrangements should be made for the erection of from one thousand to two thousand houses at the onset.

In the Village a Co-operative Goods Store should be established, supplying everything that was really necessary for the villagers at the most economic prices.

The sale of intoxicating drink should be strictly forbidden on the Estate, and, if possible. the landowner from whom the land is obtained should be tied off from allowing any licences to be held on any other portion of the adjoining land.

It is thought that the Railway Company, in consideration of the inconvenience and suffering they have inflicted on the poor, and in their own interests, might be induced to make the following advantageous arrangements :—

(1) The conveyance of each member actually living in the village to and from London at the rate of sixpence per week. Each pass should have on it the portrait of the owner, and be fastened to some article of the dress, and be available only by Workmen's Trains running early and late and during certain hours of the day, when the trains are almost empty.

(2) The conveyance of goods and parcels should be at half the ordinary rates.

It is reasonable to suppose that large landowners would gladly give one hundred acres of land in view of the immensely advanced values of the surrounding property which would immediately follow, seeing that the erection of one thousand or two thousand cottages would constitute the nucleus of a much larger Settlement.

Lastly, the rent of a four-roomed cottage must not exceed 3s. per week. Add to this the sixpenny ticket to and from London, and you have 3s. 6d., and if the company should insist on 1s., it will make 4s., for which there would be all the advantages of a comfortable cottage—of which it would be possible for the tenant to become the owner—a good garden, pleasant surroundings, and other influences promotive of the health and happiness of the family. It is hardly necessary to remark that in connection with this Village there will be perfect freedom of opinion on all matters. A glance at the ordinary homes of the poor people of this great City will at once assure us that such a village would be a veritable Paradise to them, and that were four, five, or six settlements provided at once they would not contain a tithe of the people who would throng to occupy them.

Section 3.—THE POOR MAN'S BANK.

If the love of money is the root of all evil, the want of money is the cause of an immensity of evil and trouble. The moment you begin practically to alleviate the miseries of the people, you discover that the eternal want of pence is one of their greatest difficulties. In my most sanguine moments I have never dreamed of smoothing this difficulty out of the lot of man, but it is surely no unattainable ideal to establish a Poor Man's Bank, which will extend to the lower middle class and the working population the advantages of the credit system, which is the very foundation of our boasted commerce.

It might be better that there should be no such thing as credit, that no one should lend money, and that everyone should be compelled to rely solely upon whatever ready money he may possess from day to day. But if so, let us apply the principle all round ; do not let us glory in our world-wide commerce and boast ourselves in our riches, obtained, in so many cases, by the ignoring of this principle. If it is right for a great merchant to have dealings with his banker, if it is indispensable for the due carrying on of the business of the rich men that they should have at their elbow a credit system which will from time to time accommodate them with needful advances and enable them to stand up against the pressure of sudden demands, which otherwise would wreck them, then surely the case is still stronger for providing a similar resource for the smaller men, the weaker men. At present Society is organised far too much on the principle of giving to him who hath so that he shall have more abundantly, and taking away from him who hath not even that which he hath.

If we are to really benefit the poor, we can only do so by practical measures. We have merely to look round and see the kind of advantages which wealthy men find indispensable for the due management of their business, and ask ourselves whether poor men cannot be supplied with the same opportunities. The reason why they are not is obvious. To supply the needs of the rich is a means of making yourself rich; to supply the needs of the poor will involve you in trouble so out of proportion to the profit that the game may not be worth the candle. Men go into banking and other businesses for the sake of obtaining what the American humourist said was the chief end of man in these modern times, namely, "ten per cent." To obtain a ten per cent. what will not men do? They will penetrate the bowels of the earth, explore the depths of the sea, ascend the snow-capped mountain's highest peak, or navigate the air, if they can be guaranteed a ten per cent. I do not venture to suggest that the business of a Poor Man's Bank would yield ten per cent., or even five, but I think it might be made to pay its expenses, and the resulting gain to the community would be enormous.

Ask any merchant in your acquaintance where his business would be if he had no banker, and then, when you have his answer, ask yourself whether it would not be an object worth taking some trouble to secure, to furnish the great mass of our fellow country-men, on sound business principles with the advantages of the credit system, which is found to work so beneficially for the "well-to-do" few.

Some day I hope the State may be sufficiently enlightened to take up this business itself; at present it is left in the hands of the pawnbroker and the loan agency, and a set of sharks, who cruelly prey upon the interests of the poor. The establishment of land banks, where the poor man is almost always a peasant, has been one of the features of modern legislation in Russia, Germany, and elsewhere. The institution of a Poor Man's Bank will be, I hope, before long, one of the recognised objects of our own government.

Pending that I venture to throw out a suggestion, without in any way pledging myself to add this branch of activity to the already gigantic range of operations foreshadowed in this book—Would it not be possible for some philanthropists with capital to establish on clearly defined principles a Poor Man's Bank for the making of small loans on good security, or making advances to those who are in danger of being overwhelmed by sudden financial pressure—in fact, for doing for the "little man" what all the banks do for the "big man"?

Meanwhile, should it enter into the heart of some benevolently dis-posed possessor of wealth to give the price of a racehorse, or of an

" old master," to form the nucleus of the necessary capital, I will certainly experiment in this direction.

I can anticipate the sneer of the cynic who scoffs at what he calls my glorified pawnshop. I am indifferent to his sneers. A Mont de Piété—the very name (Mount of Piety) shows that the Poor Man's Bank is regarded as anything but an objectionable institution across the Channel—might be an excellent institution in England. Owing, however, to the vested interests of the existing traders it might be impossible for the State to establish it, excepting at a ruinous expense. There would be no difficulty, however, of instituting a private Mont de Piété, which would confer an incalculable boon upon the struggling poor.

Further, I am by no means indisposed to recognise the necessity of dealing with this subject in connection with the Labour Bureau, provided that one clearly recognised principle can be acted upon. That principle is that a man shall be free to bind himself as security for the repayment of a loan, that is to pledge himself to work for his rations until such time as he has repaid capital and interest. An illustration or two will explain what I mean. Here is a carpenter who comes to our Labour shed ; he is an honest, decent man, who has by sickness or some other calamity been reduced to destitution. He has by degrees pawned one article after another to keep body and soul together, until at last he has been compelled to pawn his tools. We register him, and an employer comes along who wants a carpenter whom we can recommend. We at once suggest this man, but then arises this difficulty. He has no tools ; what are we to do ? As things are at present, the man loses the job and continues on our hands. Obviously it is most desirable in the interest of the community that the man should get his tools out of pawn ; but who is to take the responsibility of advancing the money to redeem them ? This difficulty might be met, I think, by the man entering into a legal undertaking to make over his wages to us, or such proportion of them as would be convenient to his circumstances, we in return undertaking to find him in food and shelter until such time as he has repaid the advance made. That obligation it would be the truest kindness to enforce with Rhadamantine severity. Until the man is out of debt he is not his own master. All that he can make over his actual rations and Shelter money should belong to his creditor. Of course such an arrangement might be varied indefinitely by private agreement; the repayment of instalments could be spread over a longer or shorter time, but the mainstay of the whole principle would be the execution of a legal agreement by which the man makes over the whole product of his labour to the Bank until he has repaid his debt.

Take another instance. A clerk who has been many years in a situation and has a large family, which he has brought up respectably and educated. He has every prospect of retiring in a few years upon a superannuating allowance, but is suddenly confronted by a claim often through no fault of his own, of a sum of fifty or a hundred pounds, which is quite beyond his means. He has been a careful, saving man, who has never borrowed a penny in his life, and does not know where to turn in his emergency. If he cannot raise this money he will be sold up, his family will be scattered, his situation and his prospective pension will be lost, and blank ruin will stare him in the face. Now, were he in receipt of an income of ten times the amount, he would probably have a banking account, and, in consequence, be able to secure an advance of all he needed from his banker. Why should he not be able to pledge his salary, or a portion of it, to an Institution which would enable him to pay off his debt, on terms that, while sufficiently remunerative to the bank, would not unduly embarrass him ?

At present what does the poor wretch do ? He consults his friends, who, it is quite possible, are as hard up as himself, or he applies to some loan agency, and as likely as not falls into the hands of sharpers, who indeed, let him have the money, but at interest altogether out of proportion to the risk which they run, and use the advantage which their position gives them to extort every penny he has. A great black book written within and without in letters of lamentation, mourning, and woe might be written on the dealings of these usurers with their victims in every land.

It is of little service denouncing these extortioners. They have always existed, and probably always will ; but what we can do is to circumscribe the range of their operations and the number of their victims. This can only be done by a legitimate and merciful provision for these poor creatures in their hours of desperate need, so as to prevent their falling into the hands of these remorseless wretches, who have wrecked the fortunes of thousands, and driven many a decent man to suicide or a premature grave.

There are endless ramifications of this principle, which do not need to be described here, but before leaving the subject I may allude to an evil which is a cruel reality, alas ! to a multitude of unfortunate men and women.   I refer to the working of the Hire System. The decent poor man or woman who is anxious to earn an honest penny by the use of, it may be a mangle, or a sewing-machine, a lathe, or some other indispensable instrument, and is without the few pounds necessary to buy it, must take it on the Hire System—that is to say, for the accommodation of being allowed to pay for the machine by instalments—he is charged, in addition to the full market value of his purchase, ten or twenty times the amount of what would be a fair rate of interest, and more than this if he should at any time, through misfortune, fail in his payment, the total amount already paid will be confiscated, the machine seized, and the money lost.

Here again we fall back on our analogy of what goes on in a small community where neighbours know each other. Take, for instance, when a lad who is recognised as bright, promising, honest, and industrious, who wants to make a start in life which requires some little outlay, his better-to-do neighbour will often assist him by providing the capital necessary to enable him to make a way for himself in the world. The neighbour does this because he knows the lad, because the family is at least related by ties of neighbourhood, and the honour of the lad's family is a security upon which a man may safely advance a small sum. All this would equally apply to a destitute widow, an artizan suddenly thrown out of work, an orphan family, or the like. In the large City all this kindly helpfulness disappears, and with it go all those small acts of service which are, as it were, the buffers which save men from being crushed to death against the iron walls of circumstances. We must try to replace them in some way or other if we are to get back, not to the Garden of Eden, but to the ordinary conditions of life, as they exist in a healthy, small community. No institution, it is true, can ever replace the magic bond of personal friendship, but if we have the whole mass of Society permeated in every direction by brotherly associations established for the purpose of mutual help and sympathising counsel, it is not an impossible thing to believe that we shall be able to do something to restore the missing element in modern civilisation.

Section 4.—THE POOR MAN'S LAWYER.

The moment you set about dealing with the wants of the people, you discover that many of their difficulties are not material, but moral. There never was a greater mistake than to imagine that you have only to fill a man's stomach, and clothe his back in order to secure his happiness. Man is, much more than a digestive apparatus, liable to get out of order. Hence, while it is important to remember that man has a stomach, it is also necessary to bear in mind that he has a heart, and a mind that is frequently sorely troubled by diffi- culties which, if he lived in a friendly world, would often disappear. A man, and still more a woman, stands often quite as much in need of a trusted adviser as he or she does of a dinner or a dress. Many a poor soul is miserable all the day long, and gets dragged down deeper and deeper into the depths of sin and sorrow and despair for want of a sympathising friend, who can give her advice, and make her feel that somebody in the world cares for her, and will help her if they can.

If we are to bring back the sense of brotherhood to the world, we must confront this difficulty. God, it was said in old time, setteth the desolate in families ; but somehow, in our time, the desolate wander alone in the midst of a careless and unsympathising world. " There is no one who cares for my soul. There is no creature loves me, and if I die no one will pity me," is surely one of the bitterest cries that can burst from a breaking heart. One of the secrets of the success of the Salvation Army is, that the friendless of the world find friends in it. There is not one sinner in the world— no matter how degraded and dirty he may be—whom my people will not rejoice to take by the hand and pray with, and labour for, if thereby they can but snatch him as a brand from the burning. Now, we want to make more use of this, to make the Salvation Army the nucleus of a great agency for bringing comfort and counsel to those who are at their wits' end, feeling as if in the whole world there was no one to whom they could go.

What we want to do is to exemplify to the world the family idea. " Our Father " is the keynote.   One is Our Father, then all we are brethren.   But in a family, if anyone is troubled in mind or conscience, there is no difficulty.   The daughter goes to her father, or the son to his mother, and pour out their soul's troubles, and are relieved.   If there is any serious difficulty a family council is held, and all unite their will and their resources to get matters put straight.   This is what we mean to try to get done in the New Organisation of Society for which we are labouring.   We cannot know better than God Almighty what will do good to man.   We are content to follow on His lines, and to mend the world we shall seek to restore something of the family idea to the many hundreds of thousands — ay,   millions — who   have   no   one   wiser   or   more experienced than themselves, to whom they can take their sorrows, or consult in their difficulties. .

Of course we can do this but imperfectly.   Only God can create a mother.   But Society needs a great deal of mothering, much more than it gets.   And as a child needs a mother to run to in its difficulties and troubles, to whom it can let out its little heart in confidence, so men and women, weary and worn in the battles of life, need someone to whom they can go when pressed down with a sense of wrongs suffered or done, knowing that their confidence will be preserved inviolate, and that their statements will be received with sympathy.   I propose to attempt to meet this want. I shall establish a department, over which I shall place the wisest, the pitifullest, and the most sagacious men and women whom I can find on my staff, to whom all those in trouble and perplexity shall be invited to address themselves.   It is no use saying that we love our fellow men unless we try to help them, and it is no use pretending to sympathise with the heavy burdens which darken their lives unless we try to ease them and to lighten their existence.

Insomuch as we have more practical experience of life than other men, by so much are we bound to help their inexperience, and share our talents with them.   But if we believe they are our brothers, and that One is our Father, even the God who will come to judge us hereafter for all the deeds that we have done in the body, then must we constitute, in some such imperfect way as is open to us, the parental office.   We must be willing to receive the outpourings of our struggling fellow men, to listen to the long-buried secret that has troubled the human heart, and to welcome instead of repelling those who would obey the Apostolic precept: "To confess their sins one to another." Let not that word confession scandalise any.   Confession of the most open sort ; confession on the public platform before the presence of all the man's former associates in sin has long been one of the most potent weapons by which the Salvation Army has won its victories.   That confession we have long imposed on all our converts, and it is the only confession which seems to us to be a condition of Salvation.   But this suggestion is of a different kind.   It is not imposed as a means of grace.   It is not put forward as a preliminary to the absolution which no one can pronounce but our Lord Himself.   It is merely a response on our part to one of the deepest needs and secret longings of the actual men and women who are meeting us daily in our work.   Why should they be left to brood in misery over their secret sin, when a plain straightforward talk with a man or woman selected for his or her sympathetic common-sense and spiritual experience might take the weight off their shoulders which is crushing them into dull despair ?

Not for absolution, but for sympathy and direction, do I propose to establish my Advice Bureau in definite form, for in practice it has been in existence for some time, and wonderful things have been done in the direction on which I contemplate it working.   I have no pleasure in inventing these departments.   They all entail hard work and no end of anxiety.   But if we are to represent the love of God to men, we must minister to all the wants and needs of the human heart.   Nor is it only in affairs of the heart that this Advice Bureau will be of service.   It will be quite as useful in affairs of the head.   As I conceive it, the Advice Bureau will be

THE POOR MAN'S LAWYER AND THE POOR MAN'S TRIBUNE.

There are no means in London, so far as my knowledge goes, by which the poor and needy can obtain any legal assistance in the varied oppressions and difficulties from which they must, in consequence of their poverty and associations, be continually suffering.

While the " well-to-do " classes can fall back upon skilful friends for direction, or avail themselves of the learning and experience of the legal profession, the poor man has literally no one qualified to counsel him on such matters.   In cases of sickness he can apply to the parish doctor or the great hospital, and receive an odd word or two of advice, with a bottle of physic which may or may not be of service. But if his circumstances are sick, out of order, in danger of carrying him to utter destitution, or to prison, or to the Union, he has no one to appeal to who has the willingness or the ability to help him.

Now, we want to create a Court of Counsel or Appeal, to which anyone suffering from imposition having to do with person, liberty, or property, or anything else of sufficient importance, can apply, and obtain not only advice, but practical assistance.

Among others for whom this Court would be devised is the shamefully-neglected class of Widows, of whom in the East of London there are 6,000, mostly in very destitute circumstances. In the whole of London there cannot be less than 20,000, and in England and Wales it is estimated there are 100,000, fifty thousand of whom are probably poor and friendless.

The treatment of these poor people by the nation is a crying scandal. Take the case of the average widow, even when left in comfortable circumstances. She will often be launched into a sea of perplexity, although able to avail herself of the best advice. But think of the multitudes of poor women, who, when they close their husbands' eyes, lose the only friend who knows anything about their circumstances. There may be a trifle of money or a struggling business or a little income connected with property or some other possession, all needing immediate attention, and that of a skilful sort, in order to enable the poor creature to weather the storm and avoid the vortex of utter destitution.

All we have said applies equally to orphans and friendless people generally. Nothing, however, short of a national institution could meet the necessities of all such cases. But we can do something, and in matters already referred to, such as involve loss of property, malicious prosecution, criminal and otherwise, we can render substantial assistance.

In carrying out this purpose it will be no part of our plan to encourage legal proceedings in others, or to have recourse to them ourselves. All resort to law would be avoided either in counsel or practice, unless absolutely necessary. But where manifest injustice and wrong are perpetrated, and every other method of obtaining reparation fails, we shall avail ourselves of the assistance the Law affords.

Our great hope of usefulness, however, in this Department lies in prevention. The knowledge that the oppressed poor have in us a friend able to speak for them will often prevent the injustice which cowardly and avaricious persons might otherwise inflict, and the same considerations may induce them to accord without compulsion the right of the weak and friendless.

I also calculate upon a wide sphere of usefulness in the direction of friendly arbitration and intervention. There will be at least one disinterested tribunal, however humble, to which business, domestic, or any other questions of a contentious and litigious nature can be referred without involving any serious costs.

The following incidents have been gathered from operations already undertaken in this direction, and will explain and illustrate the kind of work we contemplate, and some of the benefits that may be expected to follow from it.

About four years ago a young and delicate girl, the daughter of a pilot, came to us in great distress. Her story was that of thousands of others. She had been betrayed by a man in a good position in the West End, and was now the mother of an infant child.

Just before her confinement her seducer had taken her to his solicitors and made her sign and swear an affidavit to the effect that he was not the father of the then expected child. Upon this he gave her a few pounds in settlement of all claims upon him. The poor thing was in great poverty and distress. Through our solicitors, we immediately opened communications with the man, and after negotiations, he, to avoid further proceedings, was compelled to secure by a deed a proper allowance to his unfortunate victim for the maintenance of her child.

SHADOWED AND CAUGHT.

A—— was induced to leave a comfortable home to become the governess of the motherless children of Mr. G——, whom she found to be a kind and considerate employer. After she had been in his service some little time he proposed that she should take a trip to London. To this she very gladly consented, all the more so when he offered to take her himself to a good appointment he had secured for her. In London he seduced her, and kept her as his mistress until, tired of her, he told her to go and do as "other women did."

Instead of descending to this infamy, she procured work, and so supported herself and child in some degree of comfort, when he sought her out and again dragged her down. Another child was born, and a second time he threw her up and left her to starve. It was then she applied to our people. We hunted up the man, followed him to the country, threatened him with public exposure, and forced from him the payment to his victim of £60 down, an allowance of £1 a week, and an Insurance Policy on his life for £450 in her favour.

£60 FROM ITALY.

C. was seduced by a young Italian of good position in society, who promised to marry her, but a short time before the day fixed for the ceremony he told her urgent business called him abroad. He assured her he would return in two years and make her his wife. He wrote occasionally, and at last broke her heart by sending the news of his marriage to another, adding insult to injury by suggesting that she should come and live with his wife as her maid, offering at the same time to pay for the maintenance of the child till it was old enough to be placed in charge of the captain of one of the vessels belonging to his firm.

None of these promises were fulfilled, and C., with her mother's assistance, for a time managed to support herself and child; but the mother, worn out by age and trouble, could help her no longer, and the poor girl was driven to despair. Her case was brought before us, and we at once set to work to assist her. The Consul of the town where the seducer lived in style was communicated with. Approaches were made to the young man's father, who, to save the dishonour that would follow exposure, paid over £60. This helps to maintain the child; and the girl is in domestic service and doing well.

THE HIRE SYSTEM.

The most cruel wrongs are frequently inflicted on the very poorest persons, in connection with this method of obtaining Furniture, Sewing Machines, Mangles, or other articles. Caught by the lure of misleading advertisements, the poor are induced to purchase articles to be paid for by weekly or monthly instalments. They struggle through half the amount perhaps, at all manner of sacrifice, when some delay in the payment is made the occasion not only for seizing the goods, which they have come to regard as their own, and on which their very existence depends, but by availing themselves of some technical clause in the agreement, for robbing them in addition. In such circumstances the poor things, being utterly friendless, have to submit to these infamous extortions without remedy. Our Bureau will be open to all such.

TALLYMEN, MONEY LENDERS, AND BILLS-OF-SALEMONGERS.

Here again we have a class who prey upon the poverty of the people, inducing them to purchase things for which they have often no immediate use—anyway for which there is no real necessity—by all manner of specious promises as to easy terms of repayment.

And once having got their dupes into their power they drag them down to misery, and very often utter temporal ruin ; once in their net escape is exceedingly difficult, if not impossible. We propose to help the poor victims by this Scheme, as far as possible.

Our Bureau, we expect will be of immense service to Clergymen, Ministers of all denominations, District Visitors, Missionaries, and others who freely mix among the poor, seeing that they must be frequently appealed to for legal advice, which they are quite unable to give, and equally at a loss to obtain. We shall always be very glad to assist such.

THE DEFENCE OF UNDEFENDED PERSONS.

The conviction is gradually fixing itself upon the public mind that a not inconsiderable number of innocent persons are from time to time convicted of crimes and offences, the reason for which often is the mere inability to secure an efficient defence. Although there are several societies in London and the country dealing with the criminal classes, and more particularly with discharged prisoners, yet there does not appear to be one for the purpose of assisting unconvicted prisoners. This work we propose boldly to take up.

By this and many other ways we shall help those charged with criminal offences, who, on a most careful enquiry, might reasonably be supposed to be innocent, but who, through want of means, are unable to obtain the legal assistance, and produce the evidence necessary for an efficient defence.

We shall not pretend authoritatively to judge as to who is innocent or who is guilty, but if after full explanation and enquiry the person charged may reasonably be supposed to be innocent, and is not in a position to defend himself, then we should feel free to advise such a case, hoping thereby to save such person and his family and friends from much misery, and possibly from utter ruin.

Mr. Justice Field recently remarked :—

"For a man to assist another man who was under a criminal charge was a highly laudable and praiseworthy act. If a man was without friends, and an Englishman came forward and legitimately, and for the purpose of honestly assisting him with means to put before the Court his case, that was a highly laudable and praiseworthy act, and he should be the last man in the country to complain of any man for so doing."

These remarks are endorsed by most Judges and Magistrates, and our Advice Bureau will give practical effect to them.

In every case an attempt wiil be made to secure, not only the outward reformation, but the actual regeneration of all whom we assist.   Special attention, as has been described under the " Criminal Reform Department," will be paid to first offenders.

We shall endeavour also to assist, as far as we have ability, the Wives and Children of persons who are undergoing sentences, by endeavouring to obtain for them employment, or otherwise rendering them help.   Hundreds of this class fall into the deepest distress and demoralisation through want of friendly aid in the forlorn circumstances in which they find themselves on the conviction of relatives on whom they have been dependent for a livelihood, or for protection and direction in the ordinary affairs of life.

This Department will also be responsible for gathering intelligence, spreading information, and the general prosecution of su ch measures as are likely to lead to the much-needed beneficial changes in our Prison Management.   In short, it will seek to become the true friend and saviour of the Criminal Classes in general, and in doing so we shall desire to act in harmony with the societies at present in existence, who may be seeking for objects kindred to the Advice Bureau.

We pen the following list to give some idea of the topics on which the Advice Bureau may be consulted :—

Accidents, Claim for	Children, Custody of	Employers' Liability Act
Administration of Estates	Compensation for Injuries	Executors, Duties of
Adulteration of Food and	„ for Accident	
Drugs	„ for Defamation	Factory Act, Breach of
Agency, Questions of	„ for Loss of	Fraud, Attempted
Agreements, Disputed	Employ-	
Affiliation Cases	ment, &c.,	Goodwill, Sale of
Animals, Cruelty to	&c.	Guarantee, Forfeited
Arrest, Wrongful	Confiscation by Landlords	
Assault	Contracts, Breach of	Heir-at-Law
	Copyright, Infringement	Husbands and Wives,
Bankruptcies	of	Disputes of
Bills of Exchange	County Court Cases	
Bills of Sale		Imprisonment, False
Bonds, Forfeited	Debts	Infants, Custody of
Breach of Promise	Distress, Illegal	Intestacy, Cases of
	Divorce	
Children, Cruelty to	Ejectment Cases	Judgment Summonses.

P

Landlord and Tenant Cases

Leases, Lapses and Renewals of

Legacies, Disputed

Libel Cases

Licences

Marriage Law, Question of the

Masters' and Servants' Acts

Meeting, Right of Public

Mortgages

Negligence, Alleged

Next of Kin Wanted

Nuisances, Alleged

Partnership, The Law of

Patents, Registration and Infringement of

Pawnbrokers and their Pledges

Police Cases

Probate

Rates and Taxes

Reversionary Interests

Seduction, Cases of

Servants' Wrongful Dismissal

Sheriffs

Sureties Estreated

Tenancies, Disputed

Trade Marks, Infringement of

Trespass, Cases of

Trustees and Trusts

Wages Kept Back

Wills, Disputed and Unproved

Women, Cruelty to

Workmen, Grievances or &c., &c.

The Advice Bureau will therefore be, first of all, a place where men and women in trouble can come when they please to communicate in confidence the cause of their anxiety, with a certainty that they will receive a sympathetic hearing and the best advice.

Secondly, it will be a Poor Man's Lawyer, giving the best legal counsel as to the course to be pursued in the various circumstances with which the poor find themselves confronted.

Thirdly, it will act as a Poor Man's Tribune, and will undertake the defence of friendless prisoners supposed to be innocent, together with the resistance of illegal extortions, and the prosecution of offenders who refuse legal satisfaction for the wrongs they have committed.

Fourthly, it will act wherever it is called upon as a Court of Arbitration between litigants, where the decision will be according to equity, and the costs cut down to the lowest possible figure.

Such a Department cannot be improvised; but it is already in a fair way of development, and it can hardly fail to do great good.

Section 5.—OUR INTELLIGENCE DEPARTMENT.

An indispensable adjunct of this Scheme will be the institution of what may be called an Intelligence Department at Headquarters. Power, it has been said, belongs to the best informed, and if we are effectually to deal with the forces of social evil, we must have ready at our fingers' ends the accumulated experience and information of the whole world on this subject The collection of facts and the systematic record of them would be invaluable, rendering the results of the experiments of previous generations available for the information of our own.

At the present there is no central institution, either governmental or otherwise, in this country or any other, which charges itself with the duty of collecting and collating the ideas and conclusions on Social Economy, so far as they are likely to help the solution of the problem we have in hand.   The British Home Office has only begun to index its own papers.   The Local Government Board is in a similar condition, and, although each particular Blue Book may be admirably indexed, there is no classified index of the whole series. If this is the case with the Government, it is not likely that the innumerable private organisations which are pecking here and there at the social question should possess any systematised method for the purpose of comparing notes and storing information. This Intelligence Department, which I propose to found on a small scale at first, will have in. it the germ of vast extension which will, if adequately supported, become a kind of University, in which the accumulated experiences of the human race will be massed, digested, and rendered available to the humblest toiler in the great work of social reform. At the present moment, who is there that can produce in any of our museums and universities as much as a classified index of publications relating to one of the many heads under which I have dealt with this subject ?   Who is there among all our wise men and social reformers that can send me a list of all the best tracts upon—say, the establishment of agricultural colonies or the experiments that have been made in dealing with inebriates ; or the best plans for the construction of a working man's cottage ?

For the development of this Scheme I want an Office to begin with, in which, under the head of the varied subjects treated of in this volume, I may have arranged the condensed essence of all the best books that have been written, and the names and addresses of those whose opinions are worth having upon them, together with a note of what those opinions are, and the results of experiments which have been made in relation to them.   I want to establish a system which will enable me to use, not only the eyes and hands of Salvation Officers, but of sympathetic friends in all parts of the world, for purposes of noticing and reporting at once every social experiment of importance, any words of wisdom on the social question, whether it may be the breeding of rabbits, the organisation of an emigration service, the best method of conducting a Cottage Farm, or the best way of cooking potatoes.   There is nothing in the whole range of our operations upon which we should not be accumulating and recording the results of human experience.   What I want is to get the essence of wisdom which the wisest have gathered from the widest experience, rendered instantly available for the humblest worker in the Salvation Factory or Farm Colony, and for any other toiler in similar fields of social progress.

It can be done, and in the service of the people it ought to be done. I look for helpers in this department among those who hitherto may not have cared for the Salvation Army, but who in the seclusion of their studies and libraries will assist in the compiling of this great Index of Sociological Experiments, and who would be willing, in this form, to help in this Scheme, as Associates, for the ameliorating of the condition of the people, if in nothing else than in using their eyes and ears, and giving me the benefit of their brains as to where knowledge lies, and how it can best be utilised.   I propose to make a beginning by putting two capable men and a boy in an office, with instructions to cut out, preserve, and verify all contemporary records in the daily and weekly press that have a bearing upon any branch of our departments.   Round these two men and a boy will grow up, I confidently believe, a vast organisation of zealous unpaid workers, who will co-operate in making our Intelligence Department a great storehouse of information—a universal library where any man may learn what is the sum of human knowledge upon any branch of the subject which we have taken in hand.

Section 6.—CO-OPERATION IN GENERAL.

If anyone asked me to state in one word what seemed likely to be the key of the solution of the Social Problem I should answer unhesitatingly Co-operation. It being always understood that it is Co-operation conducted on righteous principles, and for wise and benevolent ends; otherwise Association cannot be expected to bear any more profitable fruit than Individualism. Co-operation is applied association—association for the purpose of production and distribution. Co-operation implies the voluntary combination of individuals to the attaining an object by mutual help, mutual counsel, and mutual effort. There is a great deal of idle talk in the world just now about capital, as if capital were the enemy of labour. It is quite true that there are capitalists not a few who may be regarded as the enemies, not only of labour, but of the human race; but capital itself, so far from being a natural enemy of labour, is the great object which the labourer has constantly in view. However much an agitator may denounce capital, his one great grievance is that he has not enough of it for himself. Capital, therefore, is not an evil in itself; on the contrary, it is good—so good that one of the great aims of the social reformer ought to be to facilitate its widest possible distribution among his fellow-men. It is the congestion of capital that is evil, and the labour question will never be finally solved until every labourer is his own capitalist.

All this is trite enough, and has been said a thousand times already, but, unfortunately, with the saying of it the matter ends. Co-operation has been brought into practice in relation to distribution with considerable success, but co-operation, as a means of production, has not achieved anything like the success that was anticipated. Again and again enterprises have been begun on co-operative principles which bid fair, in the opinion of the promoters, to succeed; but after one, two, three, or ten years, the enterprise which was started with such high hopes has dwindled away into either total or partial failure.

At present, many co-operative undertakings are nothing more or less than huge Joint Stock Limited Liability concerns, shares of which are held largely by working people, but not necessarily, and sometimes not at all by those who are actually employed in the so-called co-operative business. Now, why is this? Why do co-operative firms, co-operative factories, and co-operative Utopias so very often come to grief? I believe the cause is an open secret, and can be discerned by anyone who will look at the subject with an open eye.

The success of industrial concerns is largely a question of management. Management signifies government, and government implies authority, and authority is the last thing which co-operators of the Utopian order are willing to recognise as an element essential to the success of their Schemes. The co-operative institution which is governed on Parliamentary principles, with unlimited right of debate and right of obstruction, will never be able to compete successfully with institutions which are directed by a single brain wielding the united resources of a disciplined and obedient army of workers. Hence, to make co-operation a success you must superadd to the principle of consent the principle of authority ; you must invest in those to whom you entrust the management of your co-operative establishment the same liberty of action that is possessed by the owner of works on the other side of the street. There is no delusion more common among men than the belief that liberty, which is a good thing in itself, is so good as to enable those who possess it to dispense with all other good things. But as no man lives by bread alone, neither can nations or factories or shipyards exist solely upon unlimited freedom to have their own way. In co-operation we stand pretty much where the French nation stood immediately after the outburst of the Revolution. In the enthusiasm of the proclamation of the rights of man, and the repudiation of the rotten and effete *régime* of the Bourbons, the French peasants and workmen imagined that they were inaugurating the millennium when they scrawled Liberty, Equality, and Fraternity across all the churches in every city of France. They carried their principles of freedom and license to the logical ultimate, and attempted to manage their army on Parliamentary principles. It did not work ; their undisciplined levies were driven back ; disorder reigned in the Republican camp ; and the French Revolution would have been stifled in its cradle had not the instinct of the nation discerned in time the weak point in its armour. Menaced by foreign wars and intestine revolt, the Republic established an iron discipline in its army, and enforced obedience by the summary process of military execution. The liberty and the enthusiasm developed by the outburst of the long pent-up revolutionary forces supplied the motive power, but it was the discipline of the revolutionary armies, the stern, unbending obedience which was enforced in all ranks from the highest to the lowest, which created for Napoleon the admirable military instrument by which he shattered every throne in Europe and swept in triumph from Paris to Moscow.

In industrial affairs we are very much like the French Republic before it tempered its doctrine of the rights of man by the duty of obedience on the part of the soldier. We have got to introduce discipline into the industrial army, we have to superadd the principle of authority to the principle of co-operation, and so to enable the worker to profit to the full by the increased productiveness of the willing labour of men who are employed in their own workshops and on their own property. There is no need to clamour for great schemes of State Socialism. The whole thing can be done simply, economically, and speedily if only the workers will practice as much self-denial for the sake of establishing themselves as capitalists, as the Soldiers of the Salvation Army practice every year in Self Denial Week. What is the sense of never making a levy except during a strike ? Instead of calling for a shilling, or two shillings, a week in order to maintain men who are starving in idleness because of a dispute with their masters, why should there not be a levy kept up for weeks or months, by the workers, for the purpose of setting themselves up in business as masters ? There would then be no longer a capitalist owner face to face with the masses of the proletariat, but all the means of production, the plant, and all the accumulated resources of capital would really be at the disposal of labour. This will never be done, however, as long as co-operative experiments are carried on in the present archaic fashion.

Believing in co-operation as the ultimate solution, if to co-operation you can add subordination, I am disposed to attempt something in this direction in my new Social Scheme. I shall endeavour to start a Co-operative Farm on the principles of Ralahine, and base the whole of my Farm Colony on a Co-operative foundation.

In starting this little Co-operative Commonwealth, I am reminded by those who are always at a man's elbow to fill him with forebodings of ill, to look at the failures, which I have just referred to, which make up the history of the attempt to realise ideal commonwealths in this practical workaday world.   Now, I have read the history of the many attempts at co-operation that have been made to form communistic settlements in the United States, and am perfectly familiar with the sorrowful fate with which nearly all have been overtaken ; but the story of their failures does not deter me in the least, for I regard them as nothing more than warnings to avoid certain mistakes, beacons to illustrate the need of proceeding on a different tack. Broadly speaking, your experimental communities fail because your Utopias all start upon the system of equality and government by vote of the majority, and, as a necessary and unavoidable consequence, your Utopians get to loggerheads, and Utopia goes to smash. I shall avoid that rock.   The Farm Colony, like all the other departments of the Scheme, will be governed, not on the principle of counting noses, but on the exactly opposite principle of admitting no noses into the concern that are not willing to be guided by the directing brain.   It will be managed on principles which assert that the fittest ought to rule, and it will provide for the fittest being selected, and having got them at the top, will insist on universal and unquestioning obedience from those at the bottom.   If anyone does not like to work for his rations and submit to the orders of his superior Officers he can leave.   There is no compulsion on him to stay.   The world is wide, and outside the confines of our Colony and the operations of our Corps my authority does not extend.   But judging from our brief experience it is not from revolt against authority that the Scheme is destined to fail.

There cannot be a greater mistake in this world than to imagine that men object to be governed.   They like to be governed, provided that the governor has his "head screwed on right" and that he is prompt to hear and ready to see and recognise all that is vital to the interests of the commonwealth.   So far from there being an innate objection on the part of mankind to being governed, the instinct to obey is so universal that even when governments have gone blind, and deaf, and paralytic, rotten with corruption and hopelessly behind the times, they still contrive to live on.   Against a capable Government no people ever rebel, only when stupidity and incapacity have taken possession of the seat of power do insurrections break out.

SECTION 7.—A MATRIMONIAL BUREAU.

There is another direction in which something ought to be done to restore the natural advantages enjoyed by every rural community which have been destroyed by the increasing tendency of mankind to come together in huge masses. I refer to that which is after all one of the most important elements in every human life, that of marrying and giving in marriage. In the natural life of a country village all the lads and lasses grow up together, they meet together in religious associations, in daily employments, and in their amusements on the village green. They have learned their A, B, C and pothooks together, and when the time comes for pairing off they have had excellent opportunities of knowing the qualities and the defects of those whom they select as their partners in life. Everything in such a community lends itself naturally to the indispensable preliminaries of love-making, and courtships, which, however much they may be laughed at, contribute more than most things to the happiness of life. But in a great city all this is destroyed. In London at the present moment how many hundreds, nay thousands, of young men and young women, who are living in lodgings, are practically without any opportunity of making the acquaintance of each other, or of any one of the other sex! The street is no doubt the city substitute for the village green, and what a substitute it is!

It has been bitterly said by one who knew well what he was talking about, "There are thousands of young men to-day who have no right to call any woman by her Christian name, except the girls they meet plying their dreadful trade in our public thoroughfares." As long as that is the case, vice has an enormous advantage over virtue; such an abnormal social arrangement interdicts morality and places a vast premium upon prostitution. We must get back to nature if we have to cope with this ghastly evil.

There ought to be more opportunities afforded for healthy human intercourse between young men and young women, nor can Society rid itself of a great responsibility for all the wrecks of manhood and womanhood with which our streets are strewn, unless it does make some attempt to bridge this hideous chasm which yawns between the two halves of humanity. The older I grow the more absolutely am I opposed to anything that violates the fundamental law of the family. Humanity is composed of two sexes, and woe be to those who attempt to separate them into distinct bodies, making of each half one whole! It has been tried in monasteries and convents with but poor success, yet what our fervent Protestants do not seem to see is that we are reconstructing a similar false system for our young people without the safeguards and the restraints of convent walls or the sanctifying influence of religious conviction. The conditions of City life, the absence of the enforced companionship of the village and small town, the difficulty of young people finding harmless opportunities of friendly intercourse, all tends to create classes of celibates who are not chaste, and whose irregular and lawless indulgence of a universal instinct is one of the most melancholy features of the present state of society. Nay, so generally is this recognised, that one of the terms by which one of the consequences of this unnatural state of things is popularly known is "the social evil," as if all other social evils were comparatively unworthy of notice in comparison to this.

While I have been busily occupied in working out my Scheme for the registration of labour, it has occurred to me more than once, why could not something like the same plan be adopted in relation to men who want wives and women who want husbands? Marriage is with most people largely a matter of opportunity. Many a man and many a woman, who would, if they had come together, have formed a happy household, are leading at this moment miserable and solitary lives, suffering in body and in soul, in consequence of their exclusion from the natural state of matrimony. Of course, the registration of the unmarried who wish to marry would be a matter of much greater delicacy than the registration of the joiners and stone-masons who wish to obtain work. But the thing is not impossible. I have repeatedly found in my experience that many a man and many a woman would only be too glad to have a friendly hint as to where they might prosecute their attentions or from which they might receive proposals.

In connection with such an agency, if it were established—for I am not engaging to undertake this task—I am only throwing out a

possible suggestion as to the development in the direction of meeting
a much needed want, there might be added training homes for
matrimony. My heart bleeds for many a young couple whom I see
launching out into the sea of matrimony with no housewifery
experience. The young girls who leave our public elementary
schools and go out into factories have never been trained to home
duties, and yet, when taken to wife, are unreasonably expected to
fill worthily the difficult positions of the head of a household and
the mother of a family. A month spent before marriage in a
training home of housewifery would conduce much more to the
happiness of the married life than the honeymoon which
immediately follows it.

Especially is this the case with those who marry to go abroad
and settle in a distant country. I often marvel when I think of the
utter helplessness of the modern woman, compared with the handi-
ness of her grandmother. How many of our girls can even bake a
a loaf? The baker has killed out one of our fundamental
domestic arts. But if you are in the Backwoods or in the Prairie or
in the Bush, no baker's cart comes round every morning with the
new-made bread, and I have often thought with sorrow of the kind
of stuff which this poor wife must serve up to her hungry husband.
As it is with baking, so it is with washing, with milking, with
spinning, with all the arts and sciences of the household, which
were formerly taught, as a matter of course, to all the daughters
who were born in the world. Talk about woman's rights, one of
the first of woman's rights is to be trained to her trade, to be
queen of her household, and mother of her children.

Speaking of colonists leads me to the suggestion whether
something could not be done to supply, on a well-organised
system, the thousands of bachelor miners or the vast host of
unmarried males who are struggling with the wilderness on the
outskirts of civilisation, with capable wives from the overplus
of marriageable females who abound in our great towns. Woman
supplied in adequate quantities is the great moraliser of Society,
but woman doled out as she is in the Far West and the
Australian bush, in the proportion of one woman to about a dozen
men, is a fertile source of vice and crime. Here again we must
get back to nature, whose fundamental laws our social arrangements
have rudely set on one side with consequences which as usual she does
not fail to exact with remorseless severity. There have always been

born into the world and continue to be born boys and girls in fairly equal proportions, but with colonising and soldiering our men go away, leaving behind them a continually growing surplus of marriageable but unmarried spinsters, who cannot spin, and who are utterly unable to find themselves husbands. This is a wide field on the discussion of which I must not enter. I merely indicate it as one of those departments in which an intelligent philanthropy might find a great sphere for its endeavours; but it would be better not to touch it at all than to deal with it with light-hearted precipitancy and without due consideration of all the difficulties and dangers connected therewith. Obstacles, however, exist to be overcome and converted into victories. There is even a certain fascination about the difficult and dangerous, which appeals very strongly to all who know that it is the apparently insolvable difficulty which contains within its bosom the key to the problem which you are seeking to solve.

Section 8.—WHITECHAPEL-BY-THE-SEA.

In considering the various means by which some substantial improvement can be made in the condition of the toiling masses, recreation cannot be omitted. I have repeatedly had forced upon me the desirability of making it possible for them to spend a few hours occasionally by the seaside, or even at times three or four days. Notwithstanding the cheapened rates and frequent excursions, there are multitudes of the poor who, year in and out, never get beyond the crowded city, with the exception of dragging themselves and their children now and then to the parks on holidays or hot summer evenings. The majority, especially the inhabitants of the East of London, never get away from the sunless alleys and grimy streets in which they exist from year to year. It is true that a few here and there of the adult population, and a good many of the children, have a sort of annual charity excursion to Epping Forest, Hampton Court, or perhaps to the sea. But it is only the minority. The vast number, while possessed of a passionate love of the sea, which only those who have mixed with them can conceive, pass their whole lives without having once looked over its blue waters, or watched its waves breaking at their feet.

Now I am not so foolish as to dream that it is possible to make any such change in Society as will enable the poor man to take his wife and children for a fortnight's sojourn, during the oppressive summer days, to brace them up for their winter's task, although this might be as desirable in their case as in that of their more highly favoured fellow-creatures. But I would make it possible for every man, woman and child, to get, now and then, a day's refreshing change by a visit to that never-failing source of interest.

In the carrying out of this plan, we are met at the onset with a difficulty of some little magnitude, and that is the necessity of a vastly reduced charge in the cost of the journey. To do anything effective we must be able to get a man from Whitechapel or Stratford to the sea-side and back for a shilling.

Unfortunately, London is sixty miles from the sea. Suppose we take it at seventy miles. This would involve a journey of one hundred and forty miles for the small sum of 1s. Can this be done? I think it can, and done to pay the railway companies ; otherwise there is no ground to hope for this part of my Scheme ever being realised. But I think that this great boon can be granted to the poor people without the dividends being sensibly affected. I am told that the cost of haulage for an ordinary passenger train, carrying from five hundred to a thousand persons, is 2s. 7d. per mile ; a railway company could take six hundred passengers seventy miles there, and bring them seventy miles back, at a cost of £18 1s. 8d. Six hundred passengers at a shilling is £30, so that there would be a clear profit to the company of nearly £12 on the haulage, towards the payment of interest on the capital, wear and tear of line, &c. But I reckon, at a very moderate computation, that two hundred thousand persons would travel to and fro every season. An addition of £10,000 to the exchequer of a railway company is not to be despised, and this would be a mere bagatelle to the indirect profits which would follow the establishment of a settlement which must in due course necessarily become very speedily a large and active community.

This it would be necessary to bring home to the railway companies, and for the execution of this part of my Scheme I must wait till I get some manager sufficiently public-spirited to try the experiment. When such a man is found, I purpose to set at once about my Sea-Side Establishment. This will present the following special advantages, which I am quite certain will be duly appreciated by the very poorest of the London population :—

An estate of some three hundred acres would be purchased, on which buildings would be erected, calculated to meet the wants of this class of excursionists.

Refreshments would be provided at rates very similar to those charged at our London Food Depôts. There would, of course, be greater facilities in the way of rooms and accommodation generally.

Lodgings for invalids, children, and those requiring to make a short stay in the place would be supplied at the lowest prices. Beds for single men and single women could be charged at the low rate of sixpence a night, and children in proportion, while accommodation of a suitable character, on very moderate terms, could be arranged for married people.

No public-houses would be allowed within the precincts of the settlement.

A park, playground, music, boats, covered conveniences for bathing, without the expense of hiring a machine, and other arrangements for the comfort and enjoyment of the people would be provided.

The estate would form one of the Colonies of the general enterprise, and on it would be grown fruit, vegetables, flowers, and other produce for the use of the visitors, and sold at the lowest remunerative rates.   One of the first provisions for the comfort of the excursionists would be the erection of a large hall, affording ample shelter in case of unfavourable weather, and in this and other parts of the place there would be the fullest opportunity for ministers of all denominations to hold religious services in connection with any excursionists they might bring with them.

There would be shops for tradesmen, houses for residents, a museum with a panorama and stuffed whale ; boats would be let out at moderate prices, and a steamer to carry people so many miles out to sea, and so many miles back for a penny, with a possible bout of sickness, for which no extra charge would be made.

In fact the railway fares and refreshment arrangements would be on such a scale, that a husband and wife could have a 70-mile ride through the green fields, the new-mown hay, the waving grain or fruit laden orchards ; could wander for hours on the seashore, have comforting and nourishing refreshment, and be landed back at home sober, cheered and invigorated for the small sum of 3s. A couple of children under 12 might be added at 1s. 6d.—nay, a whole family, husband, wife and four children, supposing one is in arms, could have a day at the seaside, without obligation or charity, for 5s.

The gaunt, hungry inhabitants of the Slums would save up their halfpence, and come by thousands ; clergymen would find it possible to bring half the poor and needy occupants of their parishes ; schools, mothers' meetings, and philanthropic societies of all descriptions would come down wholesale ; in short, what Brighton is to the West End and middle classes, this place would be to the East End poor, nay, to the poor of the Metropolis generally, a Whitechapel-by-the-Sea.

Now this ought to be done apart from my Scheme altogether. The rich corporations which have the charge of the affairs of this great City, and the millionaires, who would never have amassed their fortunes but by the assistance of the masses, ought to say it shall be done. Suppose the Railway Companies refused to lend the great highways of which they have become the monopolists for such an undertaking without a subvention, then the necessary subvention should be forthcoming. If it could be made possible for the joyless toilers to come out of the sweater's den, or the stifling factory ; if the seamstress could leave her needle, and the mother get away from the weary round of babydom and household drudgery for a day now and then, to the cooling, invigorating, heart-stirring influences of the sea, it should be done, even if it did cost a few paltry thousands. Let the men and women who spend a little fortune every year in Continental tours, Alpine climbings, yacht excursions, and many another form of luxurious wanderings, come forward and say that it shall be possible for these crowds of their less fortunate brethren to have the opportunity of spending one day at least in the year by the sea.

CHAPTER VII.

CAN IT BE DONE, AND HOW?

Section I.—THE CREDENTIALS OF THE SALVATION ARMY.

Can this great work be done? I believe it can. And I believe
that it can be done by the Salvation Army, because it has ready
to hand an organisation of men and women, numerous enough
and zealous enough to grapple with the enormous undertaking.
The work may prove beyond our powers. But this is not so
manifest as to preclude us from wishing to make the attempt.
That in itself is a qualification which is shared by no other
organisation—at present. If we can do it we have the field entirely
to ourselves. The wealthy churches show no inclination to compete for the onerous privilege of making the experiment in this definite and practical form. Whether we have the power or not, we
have, at least, the will, the ambition to do this great thing for the
sake of our brethren, and therein lies our first credential for being
entrusted with the enterprise.

The second credential is the fact that, while using all material
means, our reliance is on the co-working power of God. We
keep our powder dry, but we trust in Jehovah. We go not
forth in our own strength to this battle, our dependence is
upon Him who can influence the heart of man. There is
no doubt that the most satisfactory method of raising a man
must be to effect such a change in his views and feelings that he
shall voluntarily abandon his evil ways, give himself to industry and
goodness in the midst of the very temptations and companionships
that before led him astray, and live a Christian life, an example in
himself of what can be done by the power of God in the very face
of the most impossible circumstances.

Q

But herein lies the great difficulty again and again referred to, men have not that force of character which will constrain them to avail themselves of the methods of deliverance.   Now our Scheme is based on the necessity of helping such.

Our third credential is the fact that we have already out of practically nothing achieved so great a measure of success that we think we may reasonably be entrusted with this further duty.   The ordinary operations of the Army have already effected most wonderful changes in the conditions of the poorest and worst.   Multitudes of slaves of vice in every form have been delivered not only from these habits, but from the destitution and misery which they ever produce.   Instances have been given.   Any number more can be produced.   Our experience, which has been almost world-wide, has ever shown that not only does the criminal become honest, the drunkard sober, the harlot chaste, but that poverty of the most abject and helpless type vanishes away.

Our fourth credential is that our Organisation alone of England's religious bodies is founded upon the principle of implicit obedience.

For Discipline I can answer.   The Salvation Army, largely recruited from among the poorest of the poor, is often reproached by its enemies on account of the severity of its rule.   It is the only religious body founded in our time that is based upon the principle of voluntary subjection to an absolute authority.   No one is bound to remain in the Army a day longer than he pleases.   While he remains there he is bound by the conditions of the Service.   The first condition of that Service is implicit, unquestioning obedience. The Salvationist is taught to obey as is the soldier on the field of battle.

From the time when the Salvation Army began to acquire strength and to grow from the grain of mustard seed until now, when its branches overshadow the whole earth, we have been constantly warned against the evils which this autocratic system would entail. Especially were we told that in a democratic age the people would never stand the establishment of what was described as a spiriual despotism.   It was contrary to the spirit of the times, it would be a stone of stumbling and a rock of offence to the masses to whom we appeal, and so forth and so forth.

But what has been the answer of accomplished facts to these predictions of theorists ?   Despite the alleged unpopularity of our discipline, perhaps because of the rigour of military authority upon which we have insisted, the Salvation Army has grown from year to year with a rapidity to which nothing in modern Christendom affords any parallel. It is only twenty-five years since it was born. It is now the largest Home and Foreign Missionary Society in the Protestant world. We have nearly 10,000 officers under our orders, a number increasing every day, every one of whom has taken service on the express condition that he or she will obey without questioning or gainsaying the orders from Headquarters. Of these, 4,600 are in Great Britain. The greatest number outside these islands, in any one country, are in the American Republic, where we have 1,018 officers, and democratic Australia, where we have 800.

Nor is the submission to our discipline a mere paper loyalty. These officers are in the field, constantly exposed to privation and ill-treatment of all kinds. A telegram from me will send any of them to the uttermost parts of the earth, will transfer them from the Slums of London to San Francisco, or despatch them to assist in opening missions in Holland, Zululand, Sweden, or South America. So far from resenting the exercise of authority, the Salvation Army rejoices to recognise it as one great secret of its success, a pillar of strength upon which all its soldiers can rely, a principle which stamps it as being different from all other religious organisations founded in our day.

With ten thousand officers, trained to obey, and trained equally to command, I do not feel that the organisation even of the disorganised, sweated, hopeless, drink-sodden denizens of darkest England is impossible. It is possible, because it has already been accomplished in the case of thousands who, before they were saved, were even such as those whose evil lot we are now attempting to deal with.

Our fifth credential is the extent and universality of the Army. What a mighty agency for working out the Scheme is found in the Army in this respect! This will be apparent when we consider that it has already stretched itself through over thirty different Countries and Colonies, with a permanent location in something like 4,000 different places, that it has either soldiers or friends sufficiently in sympathy with it to render assistance in almost every considerable population in the civilised world, and in much of the uncivilised, that it has nearly 10,000 separated officers whose training, and leisure, and history qualify them to become its enthusiastic and earnest co-workers. In fact, our whole people will hail it as the missing link in the great Scheme for the regeneration of mankind, enabling them to act out those impulses of their hearts which are ever prompting them to do good to the bodies as well as to the souls of men.

Take the meetings. With few exceptions, every one of these four thousand centres has a Hall in which, on every evening in the week and from early morning until nearly midnight on every Sabbath, services are being held; that nearly every service held indoors is preceded by one out of doors, the special purport of every one being the saving of these wretched crowds. Indeed, when this Scheme is perfected and fairly at work, every meeting and every procession will be looked upon as an advertisement of the earthly as well as the heavenly conditions of happiness. And every Barracks and Officer's quarters will become a centre where poor sinful suffering men and women may find sympathy, counsel, and practical assistance in every sorrow that can possibly come upon them, and every Officer throughout our ranks in every quarter of the globe will become a co-worker.

See how useful our people will be in the gathering in of this class. They are in touch with them. They live in the same street, work in the same shops and factories, and come in contact with them at every turn and corner of life. If they don't live amongst them, they formerly did. They know where to find them; they are their old chums, pot-house companions, and pals in crime and mischief. This class is the perpetual difficulty of a Salvationist's life. He feels that there is no help for them in the conditions in which they are at present found. They are so hopelessly weak, and their temptations are so terribly strong, that they go down before them. The Salvationist feels this when he attacks them in the tap-rooms, in the low lodging houses, or in their own desolate homes. Hence, with many, the Crusader has lost all heart. He has tried them so often. But this Scheme of taking them right away from their old haunts and temptations will put new life into him and he will gather up the poor social wrecks wholesale, pass them along, and then go and hunt for more.

Then see how useful this army of Officers and Soldiers will be for the regeneration of this festering mass of vice and crime when it is, so to speak, in our possession.

All the thousands of drunkards, and harlots, and blasphemers, and idlers have to be made over again, to be renewed in the spirit of their

minds, that is—made good. What a host of moral workers will be required to accomplish such a gigantic transformation. In the Army we
have a few thousands ready, anyway we have as many as can be
used at the outset, and the Scheme itself will go on manufacturing
more.   Look at the qualifications of these warriors for the work!

They have been trained themselves, brought into line and are
examples of the characters we want to produce.

They understand their pupils—having been dug out of the same
pit. Set a rogue to catch a rogue, they say, that is, we suppose,
a reformed rogue. Anyway, it is so with us.   These rough-and-
ready warriors will work shoulder to shoulder with them in the
same manual employment. They will engage in the task for love.
This is a substantial part of their religion, the moving instinct of
the new heavenly nature that has come upon them. They want
to spend their lives in doing good. Here will be an opportunity.

Then see how useful these Soldiers will be for distribution! Every
Salvation Officer and Soldier in every one of these 4,000 centres,
scattered through these thirty odd countries and colonies, with all
their correspondents and friends and comrades living elsewhere, will
be ever on the watch-tower looking out for homes and employments
where these rescued men and women can be fixed up to advantage,
nursed into moral vigour, picked up again on stumbling, and watched
over generally until able to travel the rough and slippery paths of
life alone.

I am, therefore, not without warrant for my confidence in the
possibility of doing great things, if the problem so long deemed
hopeless be approached with intelligence and determination on a
scale corresponding to the magnitude of the evil with which we
have to cope.

Section 2.—HOW MUCH WILL IT COST?

A considerable amount of money will be required to fairly launch this Scheme, and some income may be necessary to sustain it for a season, but, once fairly afloat, we think there is good reason to believe that in all its branches it will be self-supporting, unless its area of operation is largely extended, on which we fully rely. Of course, the cost of the effort must depend very much upon its magnitude. If anything is to be done commensurate with the extent of the evil, it will necessarily require a proportionate outlay. If it is only the drainage of a garden that is undertaken, a few pounds will meet the cost, but if it is a great dismal swamp of many miles in area, harbouring all manner of vermin, and breeding all kinds of deadly malaria, that has to be reclaimed and cultivated, a very different sum will not only be found necessary, but be deemed an economic investment.

Seeing that the country pays out something like Ten Millions per annum in Poor Law and Charitable Relief without securing any real abatement of the evil, I cannot doubt that the public will hasten to supply one-tenth of that sum. If you reckon that of the submerged tenth we have one million to deal with, this will only be one pound per head for each of those whom it is sought to benefit, or say

ONE MILLION STERLING

to give the present Scheme a fair chance of getting into practical operation.

According to the amount furnished, must necessarily be the extent of our operations. We have carefully calculated that with one hundred thousand pounds the scheme can be successfully set in motion, and that it can be kept going on an annual income of £30,000 which is about three and a-quarter per cent. on the balance of the million sterling, for which I ask as an earnest that the public intend to put its hand to this business with serious resolution; and our judgment is based, not on any mere imaginings, but upon the actual result of the experiments already made. Still it must be remembered that so vast and desirable an end cannot be even practically contemplated without a proportionate financial outlay.

Supposing, however, by the subscription of this amount the undertaking is fairly set afloat. The question may be asked, " What further funds will be required for its efficient maintenance ? " This question we proceed to answer. Let us look at the three Colonies apart, and then at some of the circumstances which apply to the whole. To begin with, there is

THE FINANCIAL ASPECT OF THE CITY COLONY.

Here there will be, of course, a considerable outlay required for the purchasing and fitting up of property, the acquisition of machinery, furniture, tools, and the necessary plant for carrying forward all these varied operations. These once acquired, no further outlay will be needed except for the necessary reparations.

The Homes for the Destitute will be nearly, if not quite, self-sustaining. The Superior Homes for both Single and Married people will not only pay for themselves, but return some interest on the amount invested, which would be devoted to the futherance of other parts of the Scheme.

The Refuges for Fallen Girls would require considerable funds to keep them going. But the public has never been slow to practically express its sympathy with this class of work.

The Criminal Homes and Prison Gate Operations would require continued help, but not a very great deal. Then, the work in the Slums is somewhat expensive. The eighty young women at present engaged in it cost on an average 12s. per week each for personal maintenance, inclusive of clothes and other little matters, and there are expenses for Halls and some little relief which cannot in anyway be avoided, bringing our present annual Slum outlay to over £4,000. But the poor people amongst whom they work, notwithstanding their extreme poverty, are already contributing over £1,000 per annum towards this amount, which income will increase. Still as by this Scheme we propose to add at once a hundred to the number already engaged, money will be required to keep this department going.

The Inebriate Home, I calculate, will maintain itself. All its inmates will have to engage in some kind of remunerative labour, and we calculate, in addition, upon receiving money with a considerable number of those availing themselves of its benefits. But to practically assist the half-million slaves of the cup we must have money not only to launch out but to keep our operations going.

The Food Depôts, once fitted up, pay their own working expenses.

The Emigration, Advice, and Inquiry Bureaux must maintain themselves or nearly so.

The Labour Shops, Anti-Sweating, and other similar operations will without question require money to make ends meet.

But on the whole, a very small sum of money, in proportion to the immense amount of work done, will enable us to accomplish a vast deal of good.

THE FARM COLONY FROM A FINANCIAL POINT OF VIEW.

Let us now turn to the Farm Colony, and consider it from a monetary standpoint. Here also a certain amount of money will have to be expended at the outset; some of the chief items of which will be the purchase of land, the erection of buildings, the supply of stock, and the production of first crops. There is an abundance of land in the market, at the present time, at very low prices.

It is rather important for the initial experiment that an estate should be obtained not too far from London, with land suitable for immediate cultivation. Such an estate would beyond question be expensive. After a time, I have no doubt, we shall be able to deal with land of almost any quality (and that in almost any part of the country), in consequence of the superabundance of labour we shall possess. There is no question if the scheme goes forward, but that estates will be required in connection with all our large towns and cities. I am not without hope that a sufficient quantity of land will be given, or, in any way, sold to us on very favourable terms.

When acquired and stocked, it is calculated that this land, if cultivated by spade husbandry, will support at least two persons per acre. The ordinary reckoning of those who have had experience with allotments gives five persons to three acres. But, even supposing that this calculation is a little too sanguine, we can still reckon a farm of 500 acres supporting, without any outside assistance, say, 750 persons. But, in this Scheme, we should have many advantages not possessed by the simple peasant, such as those resulting from combination, market gardening, and the other forms of cultivation already referred to, and thus we should want to place two or three times this number on that quantity of land.

By a combination of City and Town Colonies, there will be a market for at least a large portion of the products. At the rate of our present consumption in the London Food Depôts and Homes for the Destitute alone, at least 50 acres would be required for potatoes alone, and every additional Colonist would be an additional consumer.

There will be no rent to pay, as it is proposed to buy the land right out.   In the event of a great rush being made for the allotments spoken of, further land might be rented, with option of purchase.

Of course, the continuous change of labourers would tell against the profitableness of the undertaking. But this would be proportionally beneficial to the country, seeing that everyone who passes through the institution with credit makes one less in the helpless crowd.

The rent of Cottages and Allotments would constitute a small return, and at least pay interest on the money invested in them.

The labour spent upon the Colony would be constantly increasing its money value.   Cottages would be built, orchards planted, land enriched, factories run up, warehouses erected, while other improvements would be continually going forward.   All the labour and a large part of the material would be provided by the Colonists themselves.

It may be suggested that the workers would have to be maintained during the progress of these erections and manufactures, the cost of which would in itself amount to a considerable sum.   True, and for this the first outlay would be required.   But after this every cottage erected, every road made, in short every structure and improvement, would be a means of carrying forward the regenerating process, and in many cases it is expected will become a source of income.

As the Scheme progresses, it is not irrational to expect that Government, or some of the varied Local Authorities, will assist in the working out of a plan which, in so marked a manner, will relieve the rates and taxes of the country.

The salaries of Officers would be in keeping with those given in the Salvation Army, which are very low.

No wages would be paid to Colonists, as has been described, beyond pocket money and a trifle for extra service.

Although no permanent invalid would be knowingly taken into the Colonies, it is fair to assume that there will be a certain number, and also a considerable residuum of naturally indolent, half-witted people, incapable of improvement, left upon our hands.   Still, it is thought that with reformed habits, variety of employment, and careful oversight, such may be made to earn their own maintenance, at least, especially when it is borne in mind that unless they work, so far as they have ability, they cannot remain in the Colony.

If the Household Salvage Scheme which has been explained in Chapter II. proves the success we anticipate, there can be no question that great financial assistance will be rendered by it to the entire scheme when once the whole thing has been brought into working order.

THE FINANCIAL ASPECT OF THE COLONY OVER-SEA.

Let us now turn to the Colony Over-Sea, and regard it also from the financial standpoint. Here we must occupy ourselves chiefly with the preliminary outlay, as we could not for a moment contemplate having to find money to assist it when once fairly established. The initial expense will, no doubt, be somewhat heavy, but not beyond a reasonable amount.

The land required would probably be given, whether we go to Africa, Canada, or elsewhere ; anyway, it would be acquired on such easy terms as would be a near approach to a gift.

A considerable sum would certainly be necessary for effecting the first settlements. There would be temporary buildings to erect, land to break up and crop ; stock, farm implements, and furniture to purchase, and other similar expenses. But this would not be undertaken on a large scale, as we should rely, to some extent, on the successive batches of Colonists more or less providing for themselves, and in this respect working out their own salvation.

The amount advanced for passages, outfit money, and settlement would be repaid by instalments by the Colonists, which would in turn serve to pay the cost of conveying others to the same destination.

Passage and outfit money would, no doubt, continue to be some difficulty. £8 per head, say to Africa—£5 passage money, and £3 for the journey across the country—is a large sum when a considerable number are involved ; and I am afraid no Colony would be reached at a much lower rate. But I am not without hope that the Government might assist us in this direction.

Taking up the entire question, that is of the three Colonies, we are satisfied that the sum named will suffice to set to work an agency which will probably rescue from lives of degradation and immorality an immense number of people, and that an income of something like £30,000 will keep it afloat. But supposing that a much larger amount should be required, by operations greatly in advance of those here spoken of, which we think exceedingly probable, it is not unreasonable to expect that it will be forthcoming, seeing that caring for the poor is not only a duty of universal obligation, a root principle of all religion, but an instinct of humanity not likely to be abolished in our time. We are not opposed to charity as such, but to the mode of its administration, which, instead of permanently relieving, only demoralises and plunges the recipients lower in the mire, and so defeats its own purpose.

" What ! " I think I hear some say, " a million sterling ! how can any man out of Bedlam dream of raising such a sum ? " Stop a little ! A million may be a great deal to pay for a diamond or a palace, but it is a mere trifle compared with the sums which Britain lavishes whenever Britons are in need of deliverance if they happen to be imprisoned abroad. The King of Ashantee had captive some British subjects—not even of English birth—in 1869. John Bull despatched General Wolseley with the pick of the British army, who smashed Koffee Kalkallee, liberated the captives, and burnt Coomassie, and never winced when the bill came in for £750,000. But that was a mere trifle. When King Theodore, of Abyssinia, made captives of a couple of British representatives, Lord Napier was despatched to rescue. He marched his army to Magdala, brought back the prisoners, and left King Theodore dead. The cost of that expedition was over nine millions sterling. The Egyptian Campaign, that smashed Arabi, cost nearly five millions. The rush to Khartoum, that arrived too late to rescue General Gordon, cost at least as much. The Afghan war cost twenty-one millions sterling. Who dares then to say that Britain cannot provide a million sterling to rescue, not one or two captives, but a million, whose lot is quite as doleful as that of the prisoners of savage kings, but who are to be found, not in the land of the Soudan, or in the swamps of Ashantee, or in the Mountains of the Moon, but here at our very doors ? Don't talk to me about the impossibility of raising the million. Nothing is impossible when Britain is in earnest. All talk of impossibility only means that you don't believe that the nation cares to enter upon a serious campaign against the enemy at our gates. When John Bull goes to the wars he does not count the cost. And who dare deny that the time has fully come for a declaration of war against the Social Evils which seem to shut out God from this our world ?

Section 3.—SOME ADVANTAGES STATED.

This Scheme takes into its embrace all kinds and classes of men who may be in destitute circumstances, irrespective of their character or conduct, and charges itself with supplying at once their temporal needs; and then aims at placing them in a permanent position of comparative comfort, the only stipulation made being a willingness to work and to conform to discipline on the part of those receiving its benefit.

While at the commencement, we must impose some limits with respect to age and sickness, we hope, when fairly at work, to be able to dispense with even these restrictions, and to receive any unfortunate individual who has only his misery to recommend him and an honest desire to get out of it.

It will be seen that, in this respect, the Scheme stands head and shoulders above any plan that has ever been mooted before, seeing that nearly all the other charitable and remedial proposals more or less confess their utter inability to benefit any but what they term the " decent " working man.

This Scheme seeks out by all manner of agencies, marvellously adapted for the task, the classes whose welfare it contemplates, and, by varied measures and motives adapted to their circumstances, compels them to accept its benefits.

Our Plan contemplates nothing short of revolutionising the character of those whose faults are the reason for their destitution. We have seen that with fully fifty per cent. of these their own evil conduct is the cause of their wretchedness. To stop short with them of anything less than a real change of heart will be to invite and ensure failure. But this we are confident of effecting— anyway, in the great majority of cases, by reasonings and persuasions, concerning both earthly and heavenly advantages, by the power of man, and by the power of God.

By this Scheme any man, no matter how deeply he may have
fallen in self-respect and the esteem of all about him, may re-enter
life afresh, with the prospect of re-establishing his character when
lost, or perhaps of establishing a character for the first time, and
so obtaining an introduction to decent employment, and a claim for
admission into Society as a good citizen. While many of this crowd
are absolutely without a decent friend, others will have, on that
higher level of respectability they once occupied, some relative, or
friend, or employer, who occasionally thinks of them, and who, if
only satisfied that a real change has taken place in the prodigal, will
not only be willing, but delighted, to help them once more.

By this Scheme, we believe we shall be able to teach habits of
economy, household management, thrift, and the like. There are
numbers of men who, although suffering the direst pangs of poverty,
know little or nothing about the value of money, or the prudent use of
it ; and there are hundreds of poor women who do not know what a
decently-managed home is, and who could not make one if they had
the most ample means and tried ever so hard to accomplish it,
having never seen anything but dirt, disorder, and misery in their
domestic history. They could not cook a dinner or prepare a meal
decently if their lives were dependent on it, never having had a
chance of learning how to do it. But by this Scheme we hope to
teach these things.

By this Plan, habits of cleanliness will be created, and some
knowledge of sanitary questions in general will be imparted.

This Scheme changes the circumstances of those whose poverty
is caused by their misfortune.

To begin with, it finds work for the unemployed. This is the
chief need. The great problem that has for ages been puzzling
the brains of the political economist and philanthropist has been—
" How can we find these people work ? " No matter what other
helps are discovered, without work there is no real ground for
hope. Charity and all the other ten thousand devices are only
temporary expedients, altogether insufficient to meet the necessity.
Work, apart from the fact that it is God's method of supplying
the wants of man's composite nature, is an essential to his
well-being in every way—and on this Plan there is work,
honourable work—none of your demoralising stone-breaking,
or oakum-picking business, which tantalises and insults poverty.
Every worker will feel that he is not only occupied for his own

benefit, but that any advantage reaped over and above that which he gains himself will serve to lift some other poor wretch out of the gutter.

There would be work within the capacity of all. Every gift could be employed. For instance, take five persons on the Farm— a baker, a tailor, a shoemaker, a cook, and an agriculturist. The baker would make bread for all, the tailor garments for all, the shoemaker shoes for all, the cook would cook for all, and the agriculturist dig for all. Those who know anything which would be useful to the inhabitants of the Colony will be set to do it, and those who are ignorant of any trade or profession will be taught one.

This Scheme removes the vicious and criminal classes out of the sphere of those temptations before which they have invariably fallen in the past. Our experience goes to show that when you have, by Divine grace, or by any consideration of the advantages of a good life, or the disadvantages of a bad one, produced in a man circumstanced as those whom we have been describing, the resolution to turn over a new leaf, the temptations and difficulties he has to encounter will ordinarily master him, and undo all that has been done, if he still continues to be surrounded by old companions and allurements to sin.

Now, look at the force of the temptations this class has to fight against. What is it that leads people to do wrong—people of all classes, rich as well as poor? Not the desire to sin. They do not want to sin ; many of them do not know what sin is, but they have certain appetites or natural likings, the indulgence of which is pleasant to them, and when the desire for their unlawful gratification is aroused, regardless of the claims of God, their own highest interests, or the well-being of their fellows, they are carried away by them ; and thus all the good resolutions they have made in the past come to grief.

For instance, take the temptation which comes through the natural appetite, hunger. Here is a man who has been at a religious meeting, or received some good advice, or, perhaps, just come out of prison, with the memories of the hardships he has suffered fresh upon him, or the advice of the chaplain ringing in his ears. He has made up his mind to steal no more, but he has no means of earning a livelihood. He becomes hungry. What is he to do ? A loaf of bread tempts him, or, more likely, a gold chain which he can turn into bread. An inward struggle commences, he tries to stick to his bargain, but the hunger goes on gnawing within, and it may be there is a wife and children hungry as well as himself; so he yields to the temptation, takes the chain, and in turn the policeman takes him.

Now this man does not want to do wrong, and still less does he want to go to prison. In a sincere, dreamy way he desires to be good, and if the path were easier for him he would probably walk in it.

Again, there is the appetite for drink. That man has no thought of sinning when he takes his first glass. Much less does he want to get drunk. He may have still a vivid recollection of the unpleasant consequences that followed his last spree, but the craving is on him; the public-house is there handy; his companions press him; he yields, and falls, and, perhaps, falls to rise no more.

We might amplify, but our Scheme proposes to take the poor slave right away from the public-houses, the drink, and the companions that allure him to it, and therefore we think the chances of reformation in him are far greater.

Then think of the great boon this Scheme will be to the children, bringing them out of the slums, wretched hovels, and filthy surroundings in which they are being reared for lives of abomination of every description, into the fields, amongst the green trees and cottage homes, where they can grow up with a chance of saving both body and soul.

Think again of the change this Scheme will make for these poor creatures from the depressing, demoralising surroundings, of the sightly, filthy quarters in which they are huddled together, to the pure air and sights and sounds of the country. There is much talk about the beneficial influence of pictures, music and literature upon the multitudes. Money, like water, is being poured forth to supply such attractions in Museums, People's Palaces, and the like, for the edification and amelioration of the social condition of the masses. But " God made the country, man made the town," and if we take the people to the pictures of divine manufacture, that must be the superior plan.

Again, the Scheme is capable of illimitable application. The plaister can be made as large as the wound. The wound is certainly a very extensive one, and it seems at first sight almost ridiculous for any private enterprise to attempt dealing with it. Three millions of people, living in little short of perpetual misery have to be reached and rescued out of this terrible condition. But it can be done, and this Scheme will do it, if it is allowed a fair chance. Not all at once? True! It will take time, but it will begin to tell on the festering mass straight away. Within a measurable distance we ought to be able to take out of this black sea at least a hundred individuals a week, and there is no reason why this number should not go on increasing.

An appreciable impression on this gulf of misery would be immediately made, not only for those who are rescued from its dark waters, but for those who are left behind, seeing that for every hundred individuals removed, there is just the additional work which they performed for those who remain. It might not be much, but still it would soon count up. Supposing three carpenters are starving on employment which covered one-third of their time, if you take two away, the one left will have full employment. But it will be for the public to fix, by their contributions, the extent of our operations.

The benefits bestowed by this Scheme will be permanent in duration. It will be seen that this is no temporary expedient, such as, alas! nearly every effort hitherto made on behalf of these classes has been. Relief Works, Soup Kitchens, Enquiries into Character, Emigration Schemes, of which none will avail themselves, Charity in its hundred forms, Casual Wards, the Union, and a hundred other Nostrums may serve for the hour, but they are only at the best palliations. But this Scheme, I am bold to say, offers a substantial and permanent remedy.

In relieving one section of the community, our plan involves no interference with the well-being of any other. (See Chapter VII. Section 4, " Objections.")

This Scheme removes the all but insuperable barrier to an industrious and godly life. It means not only the leading of these lost multitudes out of the " City of Destruction " into the Canaan of plenty, but the lifting of them up to the same level of advantage with the more favoured of mankind for securing the salvation of their souls.

Look at the circumstances of hundreds and thousands of the classes of whom we are speaking. From the cradle to the grave, might not their influence in the direction of Religious Belief be summarised in one sentence, "*Atheism made easy*." Let my readers imagine theirs

to have been a similar lot. Is it not possible that, under such circumstances, they might have entertained some serious doubts as
to the existence of a benevolent God who would thus allow His
creatures to starve, or that they would have been so preoccupied with
their temporal miseries as to have no heart for any concern about
the next life?

Take a man, hungry and cold, who does not know where his
next meal is coming from; nay, who thinks it problematical whether
it will come at all. We know his thoughts will be taken up entirely
with the bread he needs for his body. What he wants is a dinner.
The interests of his soul must wait.

Take a woman with a starving family, who knows that as soon
as Monday comes round the rent must be paid, or else she and
her children must go into the street, and her little belongings be
impounded. At the present moment she is without it. Are not
her thoughts likely to wander in that direction if she slips into a
Church or Mission Hall, or Salvation Army Barracks?

I have had some experience on this subject, and have been
making observations with respect to it ever since the day I made
my first attempt to reach these starving, hungry, crowds—just
over forty-five years ago—and I am quite satisfied that these
multitudes will not be saved in their present circumstances. All
the Clergymen, Home Missionaries, Tract Distributors, Sick
Visitors, and everyone else who care about the Salvation of the
poor, may make up their minds as to that. If these people are
to believe in Jesus Christ, become the Servants of God, and
escape the miseries of the wrath to come, they must be helped
out of their present social miseries. They must be put into a
position in which they can work and eat, and have a decent room
to live and sleep in, and see something before them besides a
long, weary, monotonous, grinding round of toil, and anxious care
to keep themselves and those they love barely alive, with nothing
at the further end but the Hospital, the Union, or the Madhouse. If
Christian Workers and Philanthropists will join hands to effect this
change it will be accomplished, and the people will rise up and bless
them, and be saved; if they will not, the people will curse them
and perish.

R

Section 4.—SOME OBJECTIONS MET.

Objections must be expected. They are a necessity with regard to any Scheme that has not yet been reduced to practice, and simply signify foreseen difficulties in the working of it. We freely admit that there are abundance of difficulties in the way of working out the plan smoothly and successfully that has been laid down. But many of these we imagine will vanish when we come to close quarters, and the remainder will be surmounted by courage and patience. Should, however, this plan prove the success we predict, it must eventually revolutionise the condition of the starving sections of Society, not only in this great metropolis, but throughout the whole range of civilisation. It must therefore be worthy not only of a careful consideration but of persevering trial.

Some of these difficulties at first sight appear rather serious. Let us look at them.

Objection I.—It is suggested that the class of people for whose benefit the Scheme is designed would not avail themselves of it.

When the feast was prepared and the invitation had gone forth, it is said that the starving multitudes would not come ; that though labour was offered them in the City, or prepared for them on the Farm, they would prefer to rot in their present miseries rather than avail themselves of the benefit provided.

In order to gather the opinions of those most concerned, we consulted one evening, by a Census in our London Shelters, two hundred and fifty men out of work, and all suffering severely in consequence. We furnished a set of questions, and obtained answers from the whole. Now, it must be borne in mind that these men were under no obligation whatever to make any reply to our enquiries, much less to answer them favourably to our plan, of which they knew next to nothing.

These two hundred and fifty men were mostly in the prime of life, the greater portion of them being skilled workmen; an examination of the return papers showing that out of the entire number two hundred and seven were able to work at their trades had they the opportunity.

The number of trades naturally varied. There were some of all kinds : Engineers, Custom House Officers, Schoolmasters, Watch and Clockmakers, Sailors, and men of the different branches of the Building trade ; also a number of men who have been in business on their own account.

The average amount of wages earned by the skilled mechanics when regularly employed was 33s. per week ; the money earned by the unskilled averaged 22s. per week.

They could not be accounted lazy, as most of them, when not employed at their own trade or occupation, had proved their willingness to work by getting jobs at anything that turned up. On looking over the list we saw that one who had been a Custom House Officer had recently acted as Carpenter's Labourer ; a Type-founder had been glad to work at Chimney Sweeping ; the Schoolmaster, able to speak five languages, who in his prosperous days had owned a farm, was glad to do odd jobs as a Bricklayer's Labourer ; a Gentleman's Valet, who once earned £5 a week, had come so low down in the world that he was glad to act as Sandwich man for the magnificent sum of fourteenpence a day, and that, only as an occasional affair. In the list was a dyer and cleaner, married, with a wife and nine children, who had been able to earn 40s. a week, but had done no regular work for three years out of the last ten.

We put the following question to the entire number :—" If you were put on a farm, and set to work at anything you could do, and supplied with food, lodging, and clothing, with a view to getting you on to your feet, would you be willing to do all you could ? "

In response, the whole 250 replied in the affirmative, with one exception, and on enquiry we elicited that, being a sailor, the man was afraid he would not know how to do the work.

On being interrogated as to their willingness to grapple with the hard labour on the land, they said : " Why should we not ? Look at us. Can any plight be more miserable than ours ? "

Why not, indeed ? A glance at them would certainly make it impossible for any thoughtful person to assign a rational reason for their refusal—in rags, swarming with vermin, hungry, many of them living on scraps of food, begged or earned in the most haphazard fashion, without sufficient clothing to cover their poor gaunt limbs, most of them without a shirt.   They had to start out the next morning, uncertain which way to turn to earn a crust for dinner, or the fourpence necessary to supply them again with the humble shelter they had enjoyed that night.   The idea of their refusing employment which would supply abundantly the necessaries of life, and give the prospect of becoming, in process of time, the owner of a home, with its comforts and companionships, is beyond conception.   There is not much question that this class will not only accept the Scheme we want to set before them, but gratefully do all in their power to make it a success.

II.—*Too many would come.*

This would be very probable.   There would certainly be too many apply.   But we should be under no obligation to take more than was convenient.   The larger the number of applications the wider the field for selection, and the greater the necessity for the enlargement of our operations.

III.—*They would run away.*

It is further objected that if they did come, the monotony of the life, the strangeness of the work, together with the absence of the excitements and amusements with which they had been entertained in the cities and towns, would render their existence unbearable.   Even when left to the streets, there is an amount of life and action in the city which is very attractive.   Doubtless some would run away, but I don't think this would be a large proportion.   The change would be so great, and so palpably advantageous, that I think they would find in it ample compensation for the deprivation of any little pleasureable excitement they had left behind them in the city.   For instance, there would be—

A Sufficiency of Food.

The friendliness and sympathy of their new associates.   There would be abundance of companions of similar tastes and circumstances—not all pious.   It would be quite another matter to going single-handed on to a farm, or into a melancholy family.

Then there would be the prospect of doing well for themselves in the future, together with all the religious life, meetings, music, and freedom of the Salvation Army.

But what says our experience ?

If there be one class which is the despair of the social reformer, it is that which is variously described, but which we may term the lost women of our streets.   From the point of view of the industrial organiser, they suffer from almost every fault that human material can possess.   They are, with some exceptions, untrained to labour, demoralised by a life of debauchery, accustomed to the wildest license, emancipated from all discipline but that of starvation, given to drink, and, for the most part, impaired in health.   If, therefore, any considerable number of this class can be shown to be ready to submit themselves voluntarily to discipline, to endure deprivation of drink, and to apply themselves steadily to industry, then example will go a long way towards proving that even the worst description of humanity, when intelligently, thoroughly handled, is amenable to discipline and willing to work.   In our British Rescue Homes we receive considerably over a thousand unfortunates every year ; while all over the world, our annual average is two thousand.   The work has been in progress for three years—long enough to enable us to test very fully the capacity of the class in question to reform.

With us there is no compulsion.   If any girl wishes to remain, she remains.   If she wishes to go, she goes.   No one is detained a day or an hour longer than they choose to stay.   Yet our experience shows that, as a rule, they do not run away.   Much more restless and thoughtless and given to change, as a class, than men, the girls do not, in any considerable numbers, desert.   The average of our London Homes, for the last three years, gives only 14 per cent. as leaving on their own account, while for the year 1889 only 5 per cent.   And the entire number, who have either left or been dismissed during that year, amounts only to 13 per cent. on the whole.

IV.—*They would not work.*

Of course, to such as had for years been leading idle lives, anything like work and exhaustive labour would be very trying and wearisome, and a little patience and coaxing might be required to get them into the way of it.   Perhaps some would be hopelessly beyond salvation in this respect, and, until the time comes, if it ever does arrive, when the Government will make it a crime for an abled-bodied man to beg when there is an opportunity for him to engage in remunerative work, this class will wander abroad preying upon a generous public.   It will, however, only need to be known that any man can obtain work if he wants it, for those who have by their liberality maintained men and women in idleness to cease doing so.   And when it comes to this pass, that a man cannot eat without working, of the two evils he will choose the latter, preferring labour, however unpleasant it may be to his tastes, to actual starvation.

It must be borne in mind that the penalty of certain expulsion, which all would be given to understand would be strictly enforced would have a good influence in inducing the idlest to give work a fair trial, and once at it I should not despair of conquering the aversion altogether, and eventually being able to transform and pass these once lazy loafers as real industrious members of Society.

Again, any who have fears on this point may be encouraged by contrasting the varied and ever-changing methods of labour we should pursue, with the monotonous and uninteresting grind of many of the ordinary employments of the poor, and the circumstances by which they are surrounded.

Here, again, we fall back upon our actual experience in reclamation work.   In our Homes for Saving the Lost Women we have no difficulty of getting them to work.   The idleness of this section of the social strata has been before referred to ; it is not for a moment denied, and there can be no question, as to its being the cause of much of their poverty and distress.   But from early morn until the lights are out at night, all is a round of busy, and, to a great extent, very uninteresting labour ; while the girls have, as a human inducement, only domestic service to look forward to—of which they are in no way particularly enamoured—and yet here is no mutiny, no objection, no unwillingness to work ; in fact they appear well pleased to be kept continually at it.   Here is a report that teaches the same lesson.

A small Bookbinding Factory is worked in connection with the Rescue Homes in London.   The folders and stitchers are girls saved from the streets, but who, for various reasons, were found unsuitable for domestic service.   The Factory has solved the problem of employment for some of the most difficult cases. Two of the girls at present employed there are crippled, while one is supporting herself and two young children.

While learning the work they live in the Rescue Homes, and the few shillings they are able to earn are paid into the Home funds.   As soon as they are able to earn 12s. a week, a lodging is found for them (with Salvationists, if possible), and they are placed entirely upon their own resources.   The majority of girls working at this trade in London are living in the family, and 6s., 7s., and 8s. a week make an acceptable addition to the Home income ; but our girls who are *entirely* dependent upon their own earnings must make an average wage of 12s. a week at least. In order that they may do this we are obliged to pay higher wages than other employers. For instance, we give from 2½d. to 3d. a thousand more than the trade for binding small pamphlets ; nevertheless, after the Manager, a married man, is paid, and a man for the superintendence of the machines, a profit of about £500 has been made, and the work is *improving.* They are all paid *piecework.*

Eighteen women are supporting themselves in this way at present, and conducting themselves most admirably. One of their number acts as forewoman,. and conducts the Prayer Meeting at 12.30, the Two-minutes' Prayer after meals, etc. Their continuance in the factory is subject to their good behaviour—both at home as well as at work. *In one instance only have we had any trouble at all, and in this solitary case the girl was so penitent she was forgiven, and has done well ever since.* I think that, without exception, they are Salvation Soldiers, and will be found at nearly every meeting on the Sabbath, etc. The binding of Salvation Army publications—"The Deliverer," "All the World," the Penny Song Books, etc., almost keep us going. A little outside work for the end of the months is taken, but we are not able to make any profit generally, it is so badly paid.

It will be seen that this is a miniature factory, but still it is a factory, and worked on principles that will admit of illimitable extension, and may, I think, be justly regarded as an encouragement and an exemplification of what may be accomplished in endless variations.

V.—*Again, it is objected that the class whose benefit we contemplate would not have physical ability to work on a farm, or in the open air.*

How, it is asked, would tailors, clerks, weavers, seamstresses and the destitute people, born and reared in the slums and poverty-hovels of the towns and cities, do farm or any other work that has to do with the land ? The employment in the open air, with exposure to every kind of weather which accompanies it, would, it is said, kill them off right away.

We reply, that the division of labour before described would render it as unnecessary as it would be undesirable and uneconomical, to put many of these people to dig or to plant. Neither is it any part of our plan to do so. On our Scheme we have shown how each one would be appointed to that kind of work for which his previous knowledge and experience and strength best adapted him.

Moreover, there can be no possible comparison between the conditions of health enjoyed by men and women wandering about homeless, sleeping in the streets or in the fever-haunted lodging-houses, or living huddled up in a single room, and toiling twelve and fourteen hours in a sweater's den, and living in comparative comfort in well-warmed and ventilated houses, situated in the open country, with abundance of good, healthy food.

Take a man or a woman out into the fresh air, give them proper exercise, and substantial food. Supply them with a comfortable home, cheerful companions, and a fair prospect of reaching a position of independence in this or some other land, and a complete renewal of health and careful increase of vigour will, we expect, be one of the first great benefits that will ensue.

VI.—*It is objected that we should be left with a considerable residuum of half-witted, helpless people.*

Doubtless this would be a real difficulty, and we should have to prepare for it. We certainly, at the outset, should have to guard against too many of this class being left upon our hands, although we should not be compelled to keep anyone. It would, however, be painful to have to send them back to the dreadful life from which we had rescued them. Still, however, this would not be so ruinous a risk, looked at financially, as some would imagine. We could, we think, maintain them for 4s. per week, and they would be very weak indeed in body, and very wanting in mental, strength if they were not able to earn that amount in some one of the many forms of employment which the Colony would open up.

VII.—*Again, it will be objected that some efforts of a similar character have failed. For instance, co-operative enterprises in farming have not succeeded.*

True, but so far as I can ascertain, nothing of the character I am describing has ever been attempted. A large number of Socialistic communities have been established and come to grief in the United States, in Germany, and elsewhere, but they have all, both in principle and practice, strikingly differed from what we are proposing here. Take one particular alone, the great bulk of these societies have not only been fashioned without any regard to the principles of Christianity, but, in the vast majority of instances, have been in direct opposition to them; and the only communities based on co-operative principles that have survived the first few months of their existence have been based upon Christian truth. If not absolute successes, there have been

some very remarkable results obtained by efforts partaking somewhat of the nature of the one I am setting forth. (See that of
Ralahine, described in Appendix.)

VIII.—*It is further objected that it would be impossible to maintain
order and enforce good discipline amongst this class of people.*

We are of just the opposite opinion. We think that it would—
nay, we are certain of it, and we speak as those who have had
considerable experience in dealing with the lower classes of
Society. We have already dealt with this difficulty. We may say
further—

That we do not propose to commence with a thousand people
in a wild, untamed state, either at home or abroad. To the
Colony Over-Sea we should send none but those who have had a
long period of training in this country. The bulk of those sent
to the Provincial Farm would have had some sort of trial in the
different City Establishments. We should only draft them on to
the Estate in small numbers, as we were prepared to deal with
them, and I am quite satisfied that without the legal methods of
maintaining order that are acted upon so freely in workhouses
and other similar institutions, we should have as perfect obedience
to Law, as great respect for authority, and as strong a spirit of
kindness pervading all ranks throughout the whole of the community as could be found in any other institution in the land.

It will be borne in mind that our Army system of government
largely prepares us, if it does not qualify us, for this task. Anyway,
it gives us a good start. All our people are trained in habits of
obedience, and all our Officers are educated in the exercise of
authority. The Officers throughout the Colony would be almost
exclusively recruited from the ranks of the Army, and everyone of
them would go to the work, both theoretically and practically,
familiar with those principles which are the essence of good
discipline.

Then we can argue, and that very forcibly, from the actual
experience we have already had in dealing with this class. Take
our experience in the Army itself. Look at the order of our Soldiers.
Here are men and women, who have no temporal interest whatever
at stake, receiving no remuneration, often sacrificing their earthly
interests by their union with us, and yet see how they fall into line,
and obey orders in the promptest manner, even when such orders
go right in the teeth of their temporal interests.

"Yes," it will be replied by some, "this is all very excellent so far as it relates to those who are altogether of your own way of thinking. You can command them as you please, and they will obey, but what proof have you given of your ability to control and discipline those who are not of your way of thinking?

"You can do that with your Salvationists because they are saved, as you call it. When men are born again you can do anything with them. But unless you convert all the denizens of Darkest England, what chance is there that they will be docile to your discipline? If they were soundly saved no doubt something might be done. But they are not saved, soundly or otherwise; they are lost. What reason have you for believing that they will be amenable to discipline?"

I admit the force of this objection; but I have an answer, and an answer which seems to me complete. Discipline, and that of the most merciless description, is enforced upon multitudes of these people even now. Nothing that the most authoritative organisation of industry could devise in the excess of absolute power, could for a moment compare with the slavery enforced to-day in the dens of the sweater. It is not a choice between liberty and discipline that confronts these unfortunates, but between discipline mercilessly enforced by starvation and inspired by futile greed, and discipline accompanied with regular rations and administered solely for their own benefit. What liberty is there for the tailors who have to sew for sixteen to twenty hours a day, in a pest-hole, in order to earn ten shillings a week? There is no discipline so brutal as that of the sweater; there is no slavery so relentless as that from which we seek to deliver the victims. Compared with their normal condition of existence, the most rigorous discipline which would be needed to secure the complete success of any new individual organisation would be an escape from slavery into freedom.

You may reply, "that it might be so, if people understood their own interest. But as a matter of fact they do not understand it, and that they will never have sufficient far-sightedness to appreciate the advantages that are offered them."

To this I answer, that here also I do not speak from theory. I lay before you the ascertained results of years of experience. More than two years ago, moved by the misery and despair of the unemployed, I opened the Food and Shelter Depôts in London already described. Here are a large number of men every night, many of them of the lowest type of casuals who crawl about the streets, a certain proportion criminals, and about as difficult a class to manage as I should think could be got together, and while there will be 200 of them in a single building night after night, from the first opening of the doors in the evening until the last man has departed in the morning, there shall scarcely be a word of dissatisfaction ; anyway, nothing in the shape of angry temper or bad language.   No policemen are required ; indeed two or three nights' experience will be sufficient to turn the regular frequenters of the place of their own free will into Officers of Order, glad not only to keep the regulations of the place, but to enforce its discipline upon others.

Again, every Colonist, whether in the City or elsewhere, would know that those who took the interests of the Colony to heart, were loyal to its authority and principles, and laboured industriously in promoting its interests, would be rewarded accordingly by promotion to positions of influence and authority, which would also carry with them temporal advantages, present and prospective.

But one of our main hopes would be in the apprehension by the Colonists of the fact that all our efforts were put forth on their behalf.   Every man and woman on the place would know that this enterprise was begun and carried on solely for their benefit, and that of the other members of their class, and that only their own good behaviour and co-operation would ensure their reaping a personal share in such benefit.   Still our expectations would be largely based on the creation of a spirit of unselfish interest in the community.

IX. *Again, it is objected that the Scheme is too vast to be attempted by voluntary enterprise; it ought to be taken up and carried out by the Government itself.*

Perhaps so, but there is no very near probability of Government undertaking it, and we are not quite sure whether such an attempt would prove a success if it were made.   But seeing that neither Governments, nor Society, nor individuals have stood forward to undertake what God has made appear to us to be so vitally important a work, and as He has given us the willingness, and in many important senses the ability, we are prepared, if the financial help is furnished, to make a determined effort, not only to undertake but to carry it forward to a triumphant success.

X.—*It is objected that the classes we seek to benefit are too ignorant and depraved for Christian effort, or for effort of any kind, to reach and reform.*—

Look at the tramps, the drunkards, the harlots, the criminals. How confirmed they are in their idle and vicious habits. It will be said, indeed has been already said by those with whom I have conversed, that I don't know them; which statement cannot, I think, be maintained, for if I don't know them, who does?

I admit, however, that thousands of this class are very far gone from every sentiment, principle and practice of right conduct. But I argue that these poor people cannot be much more unfavourable subjects for the work of regeneration than are many of the savages and heathen tribes, in the conversion of whom Christians universally believe; for whom they beg large sums of money, and to whom they send their best and bravest people.

These poor people are certainly embraced in the Divine plan of mercy. To their class, the Saviour especially gave His attention when he was on the earth, and for them He most certainly died on the Cross.

Some of the best examples of Christian faith and practice, and some of the most successful workers for the benefit of mankind, have sprung from this class, of which we have instances recorded in the Bible, and any number in the history of the Church and of the Salvation Army.

It may be objected that while this Scheme would undoubtedly assist one class of the community by making steady, industrious workmen, it must thereby injure another class by introducing so many new hands into the labour market, already so seriously overstocked.

To this we reply that there is certainly an appearance of force in this objection; but it has, I think, been already answered in the foregoing pages. Further, if the increase of workers, which this Scheme will certainly bring about, was the beginning and the end of it, it would certainly present a somewhat serious aspect. But, even on that supposition, I don't see how the skilled worker could leave his brothers to rot in their present wretchedness, though their rescue should involve the sharing of a portion of his wages.

(1) But there is no such danger, seeing that the number of extra hands thrown on the British Labour Market must be necessarily inconsiderable.

(2) The increased production of food in our Farm and Colonial operations must indirectly benefit the working man.

(3) The taking out of the labour market of a large number of individuals who at present have only partial work, while benefiting them, must of necessity afford increased labour to those left behind.

(4) While every poor workless individual made into a wage earner will of necessity have increased requirements in proportion. For instance, the drunkard who has had to manage with a few bricks, a soap box, and a bundle of rags, will want a chair, a table, a bed, and at least the other necessary adjuncts to a furnished home, however sparely fitted up it may be.

There is no question but that when our Colonisation Scheme is fairly afloat it will drain off, not only many of those who are in the morass, but a large number who are on the verge of it. Nay, even artisans, earning what are considered good wages, will be drawn by the desire to improve their circumstances, or to raise their children under more favourable surroundings, or from still nobler motives, to leave the old country. Then it is expected that the agricultural labourer and the village artisan, who are ever migrating to the great towns and cities, will give the preference to the Colony Over-Sea, and so prevent that accumulation of cheap labour which is considered to interfere so materially with the maintenance of a high wages standard.

Section 5.—RECAPITULATION.

I have now passed in review the leading features of the Scheme, which I put forward as one that is calculated to considerably contribute to the amelioration of the condition of the lowest stratum of our Society. It in no way professes to be complete in all its details. Anyone may at any point lay his finger on this, that, or the other feature of the Scheme, and show some void that must be filled in if it is to work with effect. There is one thing, however, that can be safely said in excuse for the shortcomings of the Scheme, and that is that if you wait until you get an ideally perfect plan you will have to wait until the Millennium, and then you will not need it. My suggestions, crude though they may be, have, nevertheless, one element that will in time supply all deficiencies. There is life in them, with life there is the promise and power of adaptation to all the innumerable and varying circumstances of the class with which we have to deal. Where there is life there is infinite power of adjustment. This is no cast-iron Scheme, forged in a single brain and then set up as a standard to which all must conform. It is a sturdy plant, which has its roots deep down in the nature and circumstances of men. Nay, I believe in the very heart of God Himself. It has already grown much, and will, if duly nurtured and tended, grow still further, until from it, as from the grain of mustard-seed in the parable, there shall spring up a great tree whose branches shall overshadow all the earth.

Once more let me say, I claim no patent rights in any part of this Scheme. Indeed, I do not know what in it is original and what is not. Since formulating some of the plans, which I had thought were new under the sun, I have discovered that they have been already tried in different parts of the world, and that with great promise. It may be so with others, and in this I rejoice. I plead for no exclusiveness. The question is much too serious for such fooling as that. Here are millions of our fellow-creatures perishing amidst the breakers of the sea of life, dashed to pieces on sharp rocks, sucked under by eddying whirlpools, suffo-cated even when they think they have reached land by treacherous quicksands ; to save them from this imminent destruction I suggest that these things should be done. If you have any better plan than mine for effecting this purpose, in God's name bring it to the light and get it carried out quickly. If you have not, then lend me a hand with mine, as I would be only too glad to lend you a hand with yours if it had in it greater promise of successful action than mine.

In a Scheme for the working out of social salvation the great, the only, test that is worth anything is the success with which they attain the object for which they are devised. An ugly old tub of a boat that will land a shipwrecked sailor safe on the beach is worth more to him than the finest yacht that ever left a slip-way incapable of effecting the same object. The superfine votaries of culture may recoil in disgust from the rough-and-ready suggestions which I have made for dealing with the Sunken Tenth, but mere recoiling is no solution. If the cultured and the respectable and the orthodox and the established dignitaries and conventionalities of Society pass by on the other side we cannot follow their example. We may not be priests and Levites, but we can at least play the part of the Good Samaritan. The man who went down to Jericho and fell among thieves was probably a very improvident, reckless individual, who ought to have known better than to go roaming alone through defiles haunted by banditti, whom he even led into temptation by the careless way in which he exposed himself and his goods to their avaricious gaze. It was, no doubt, largely his own fault that he lay there bruised and senseless, and ready to perish, just as it is largely the fault of those whom we seek to help that they lie in the helpless plight in which we find them. But for all that, let us bind up their wounds with such balm as we can procure, and, setting them on our ass, let us take them to our Colony, where they may have time to recover, and once more set forth on the journey of life.

And now, having said this much by way of reply to some of my critics, I will recapitulate the salient features of the Scheme. I laid down at the beginning certain points to be kept in view as embodying those invariable laws or principles of political economy, without due regard to which no Scheme can hope for even a chance of success. Subject to these conditions, I think my Scheme will pass muster. It is large enough to cope with the evils that will confront us ; it is practicable, for it is already in course of application, and it is capable of indefinite expansion. But it would be better to pass the whole Scheme in its more salient features in review once more.

The Scheme will seek to convey benefit to the destitute classes in various ways altogether apart from their entering the Colonies. Men and women may be very poor and in very great sorrow, nay, on the verge of actual starvation, and yet be so circumstanced as to be unable to enrol themselves in the Colonial ranks. To these our cheap Food Depôts, our Advice Bureau, Labour Shops, and other agencies will prove an unspeakable boon, and will be likely by such temporary assistance to help them out of the deep gulf in which they are struggling. Those who need permanent assistance will be passed on to the City Colony, and taken directly under our control. Here they will be employed as before described. Many will be sent off to friends ; work will be found for others in the City or elsewhere, while the great bulk, after reasonable testing as to their sincerity and willingness to assist in their own salvation, will be sent on to the Farm Colonies, where the same process of reformation and training will be continued, and unless employment is otherwise obtained they will then be passed on to the Over-Sea Colony.

All in circumstances of destitution, vice, or criminality will receive casual assistance or be taken into the Colony, on the sole conditions of their being anxious for deliverance, and willing to work for it, and to conform to discipline, altogether irrespective of character, ability, religious opinions, or anything else.

No benefit will be conferred upon any individual except under extraordinary circumstances, without some return being made in labour. Even where relatives and friends supply money to the Colonists, the latter must take their share of work with their comrades. We shall not have room for a single idler throughout all our borders.

The labour allotted to each individual will be chosen in view of his past employment or ability. Those who have any knowledge of agriculture will naturally be put to work on the land ; the shoemaker will make shoes, the weaver cloth, and so on. And when there is no knowledge of any handicraft, the aptitude of the individual and the necessities of the hour will suggest the sort of work it would be most profitable for such an one to learn.

Work of all descriptions will be executed as far as possible by hand labour. The present rage for machinery has tended to produce much destitution by supplanting hand labour so exclusively that the rush has been from the human to the machine. We want, as far as is practicable, to travel back from the machine to the human.

Each member of the Colony would receive food, clothing, lodging, medicine, and all necessary care in case of sickness.

No wages would be paid, except a trifle by way of encouragement for good behaviour and industry, or to those occupying positions of trust, part of which will be saved in view of exigencies in our Colonial Bank, and the remainder used for pocket money.

The whole Scheme of the three Colonies will for all practical purposes be regarded as one ; hence the training will have in view the qualification of the Colonists for ultimately earning their livelihood in the world altogether independently of our assistance, or, failing this, fit them for taking some permanent work within our borders either at home or abroad.

Another result of this unity of the Town and Country Colonies will be the removal of one of the difficulties ever connected with the disposal of the products of unemployed labour. The food from the Farm would be consumed by the City, while many of the things manufactured in the City would be consumed on the Farm.

The continued effort of all concerned in the reformation of these people will be to inspire and cultivate those habits, the want of which has been so largely the cause of the destitution and vice of the past.

Strict discipline, involving careful and continuous oversight, would be necessary to the maintenance of order amongst so large a number of people, many of whom had hitherto lived a wild and licentious life. Our chief reliance in this respect would be upon the spirit of mutual interest that would prevail.

The entire Colony would probably be divided into sections, each under the supervision of a sergeant—one of themselves—working side by side with them, yet responsible for the behaviour of all.

The chief Officers of the Colony would be individuals who had given themselves to the work, not for a livelihood, but from a desire to be useful to the suffering poor. They would be selected

s

at the outset from the Army, and that on the ground of their possessing certain capabilities for the position, such as knowledge of the particular kind of work they had to superintend, or their being good disciplinarians and having the faculty for controlling men and being themselves influenced by a spirit of love. Ultimately the Officers, we have no doubt, would be, as is the case in all our other operations, men and women raised up from the Colonists themselves, and who will consequently, possess some special qualifications for dealing with those they have to superintend.

The Colonists will be divided into two classes: the 1st, the class which receives no wages will consist of :—

(*a*) The new arrivals, whose ability, character, and habits are as yet unknown.

(*b*) The less capable in strength, mental calibre, or other capacity.

(*c*) The indolent, and those whose conduct and character appeared doubtful. These would remain in this class, until sufficiently improved for advancement, or are pronounced so hopeless as to justify expulsion.

The 2nd class would have a small extra allowance, a part of which would be given to the workers for private use, and a part reserved for future contingencies, the payment of travelling expenses, etc. From this class we should obtain our petty officers, send out hired labourers, emigrants, etc., etc.

Such is the Scheme as I have conceived it. Intelligently applied, and resolutely persevered in, I cannot doubt that it will produce a great and salutary change in the condition of many of the most hopeless of our fellow countrymen. Nor is it only our fellow countrymen to whom it is capable of application. In its salient features, with such alterations as are necessary, owing to differences of climate and of race, it is capable of adoption in every city in the world, for it is an attempt to restore to the masses of humanity that are crowded together in cities, the human and natural elements of life which they possessed when they lived in the smaller unit of the village or the market town. Of the extent of the need there can be no question. It is, perhaps, greatest in London, where the masses of population are denser than those of any other city ; but it exists equally in the chief centres of population in the new Englands that have sprung up beyond the sea, as well as in the larger cities of Europe. It is a remarkable fact that up to the present moment the most eager welcome that has been extended to this Scheme reaches us from Melbourne, where our officers have been compelled to begin operations by the pressure of public opinion and in compliance with the urgent entreaties of the Government on one side and the leaders of the working classes on the other before the plan had been elaborated, or instructions could be sent out for their guidance.

It is rather strange to hear of distress reaching starvation point in a city like Melbourne, the capital of a great new country which teems with natural wealth of every kind. But Melbourne, too, has its unemployed, and in no city in the Empire have we been more successful in dealing with the social problem than in the capital of Victoria. The Australian papers for some weeks back have been filled with reports of the dealings of the Salvation Army with the unemployed of Melbourne. This was before the great Strike. The Government of Victoria practically threw upon our officers the task of dealing with the unemployed. The subject was debated in the House of Assembly, and at the close of the debate a subscription was taken up by one of those who had been our most strenuous opponents, and a sum of £400 was handed over to our officers to dispense in keeping the starving from perishing. Our people have found situations for no fewer than 1,776 persons, and are dispensing meals at the rate of 700 a day. The Government of Victoria has long been taking the lead in recognising the secular uses of the Salvation Army. The following letter addressed by the Minister of the Interior to the Officer charged with the oversight of this part of our operations, indicates the estimation in which we are held :—

Government of Victoria, Chief Secretary's Office,
Melbourne.

July 4th, 1889.

Superintendent Salvation Army Rescue Work.

Sir,—In compliance with your request for a letter of introduction which may be of use to you in England, I have much pleasure in stating from reports furnished by Officers of my Department, I am convinced that the work you have been engaged on during the past six years has been of material advantage to the community. You have rescued from crime some who, but for the counsel and assistance rendered them, might have been a permanent tax upon the State, and you have restrained from further criminal courses others who had already suffered legal punishment for their misdeeds. It has given me pleasure to obtain from the Executive Council authority for you to apprehend children found in Brothels, and to take charge of such children after formal committal. Of the great value of this branch of your work there can be no question. It is evident that the attendance of yourself and your Officers at the police-courts and lock-ups has been attended with beneficial results, and your invitation to our largest jails has been highly approved by the head of the Department. Generally speaking, I may say that your policy and procedure have been commended by the Chief Officers of the Government of this Colony, who have observed your work.

I have the honour to be, Sir,

Your obedient Servant,

(Signed)   ALFRED DEAKIN.

The Victorian Parliament has voted an annual grant to our funds, not as a religious endowment, but in recognition of the service which we render in the reclamation of criminals, and what may be called, if I may use a word which has been so depraved by Continental abuse, the moral police of the city. Our Officer in Melbourne has an official position which opens to him almost every State institution and all the haunts of vice where it may be necessary for him to make his way in the search for girls that have been decoyed from home or who have fallen into evil courses.

It is in Victoria also that a system prevails of handing over first offenders to the care of the Salvation Army Officers, placing them in recognizance to come up when called for An Officer of the Army attends at every Police Court, and the Prison Brigade is always on guard at the gaol doors when the prisoners are discharged. Our Officers also have free access to the prisons, where they can conduct services and labour with the inmates for their Salvation. As Victoria is probably the most democratic of our colonies, and the one in which the working-class has supreme control, the extent to which it has by its government recognised the value of our operations is sufficient to indicate that we have nothing to fear from the opposition of the democracy. In the neighbouring colony of New South Wales a lady has already given us a farm of three hundred acres fully stocked, on which to begin operations with a Farm Colony, and there seems some prospect that the Scheme will get itself into active shape at the other end of the world before it is set agoing in London. The eager welcome which has thus forced the initiative upon our Officers in Melbourne tends to encourage the expectation that the Scheme will be regarded as no quack application, but will be generally taken up and quickly set in operation all round the world.

CHAPTER VIII.

A PRACTICAL CONCLUSION.

Throughout this book I have more constantly used the first personal pronoun than ever before in anything I have written. I have done this deliberately, not from egotism, but in order to make it more clearly manifest that here is a definite proposal made by an individual who is prepared, if the means are furnished him, to carry it out. At the same time I want it to be clearly understood that it is not in my own strength, nor at my own charge, that I purpose to embark upon this great undertaking. Unless God wills that I should work out the idea of which I believe He has given me the conception, nothing can come of any attempt at its execution but confusion, disaster, and disappointment. But if it be His will—and whether it is or not, visible and manifest tokens will soon be forth-coming—who is there that can stand against it? Trusting in Him for guidance, encouragement, and support, I propose at once to enter upon this formidable campaign.

I do not run without being called. I do not press forward to fill this breach without being urgently pushed from behind. Whether or not, I am called of God, as well as by the agonising cries of suffering men and women and children, He will make plain to me, and to us all; for as Gideon looked for a sign before he, at the bidding of the heavenly messenger, undertook the leading of the chosen people against the hosts of Midian, even so do I look for a sign. Gideon's sign was arbitrary. He selected it. He dictated his own terms; and out of compassion for his halting faith, a sign was given to him, and that twice over. First, his fleece was dry when all the country round was drenched with dew; and, secondly, his fleece was drenched with dew when all the country round was dry.

The sign for which I ask to embolden me to go forwards is single, not double. It is necessary and not arbitrary, and it is one which the veriest sceptic or the most cynical materialist will recognise as sufficient. If I am to work out the Scheme I have outlined in this book, I must have ample means for doing so. How much would be required to establish this Plan of Campaign in all its fulness, overshadowing all the land with its branches laden with all manner of pleasant fruit, I cannot even venture to form a conception. But I have a definite idea as to how much would be required to set it fairly in operation.

Why do I talk about commencing? We have already begun, and that with considerable effect. Our hand has been forced by circumstances. The mere rumour of our undertaking reaching the Antipodes, as before described, called forth such a demonstration of approval that my Officers there were compelled to begin action without waiting orders from home. In this country we have been working on the verge of the deadly morass for some years gone by, and not without marvellous effect. We have our Shelters, our Labour Bureau, our Factory, our Inquiry Officers, our Rescue Homes, our Slum Sisters, and other kindred agencies, all in good going order. The sphere of these operations may be a limited one; still, what we have done already is ample proof that when I propose to do much more I am not speaking without my book; and though the sign I ask for may not be given, I shall go struggling forward on the same lines; still, to seriously take in hand the work which I have sketched out—to establish this triple Colony, with all its affiliated agencies, I must have, at least, a hundred thousand pounds.

A hundred thousand pounds! That is the dew on my fleece. It is not much considering the money that is raised by my poor people for the work of the Salvation Army. The proceeds of the Self-Denial Week alone last year brought us in £20,000. This year it will not fall short of £25,000. If our poor people can do so much out of their poverty, I do not think I am making an extravagant demand when I ask that out of the millions of the wealth of the world I raise, as a first instalment, a hundred thousand pounds, and say that I cannot consider myself effectually called to undertake this work unless it is forthcoming.

It is in no spirit of dictation or arrogance that I ask the sign. It is a necessity. Even Moses could not have taken the Children of Israel dry-shod through the Red Sea unless the waves had divided.

That was the sign which marked out his duty, aided his faith, and determined his action. The sign which I seek is somewhat similar. Money is not everything. It is not by any means the main thing. Midas, with all his millions, could no more do the work than he could win the battle of Waterloo, or hold the Pass of Thermopylæ. But the millions of Midas are capable of accomplishing great and mighty things, if they be sent about doing good under the direction of Divine wisdom and Christ-like love.

How hardly shall they that have riches enter into the Kingdom of Heaven! It is easier to make a hundred poor men sacrifice their lives than it is to induce one rich man to sacrifice his fortune, or even a portion of it, to a cause in which, in his half-hearted fashion, he seems to believe. When I look over the roll of men and women who have given up friends, parents, home prospects, and everything they possess in order to walk bare-footed beneath a burning sun in distant India, to live on a handful of rice, and die in the midst of the dark heathen for God and the Salvation Army, I sometimes marvel how it is that they should be so eager to give up all, even life itself, in a cause which has not power enough in it to induce any reasonable number of wealthy men to give to it the mere superfluities and luxuries of their existence. From those to whom much is given much is expected ; but, alas, alas, how little is realised! It is still the widow who casts her all into the Lord's treasury—the wealthy deem it a preposterous suggestion when we allude to the Lord's tithe, and count it boredom when we ask only for the crumbs that fall from their tables.

Those who have followed me thus far will decide for themselves to what extent they ought to help me to carry out this Project, or whether they ought to help me at all. I do not think that any sectarian differences or religious feelings whatever ought to be imported into this question. Supposing you do not like my Salvationism, surely it is better for these miserable, wretched crowds to have food to eat, clothes to wear, and a home in which to lay their weary bones after their day's toil is done, even though the change is accompanied by some peculiar religious notions and practices, than it would be for them to be hungry, and naked, and homeless, and possess no religion at all. It must be infinitely preferable that they should speak the truth, and be virtuous, industrious, and contented, even if they do pray to God, sing Psalms, and go about with red jerseys, fanatically, as you call it, "seeking for the millennium "—than that they should remain thieves or harlots, with no belief in God at all, a burden to the Municipality, a curse to Society, and a danger to the State.

That you do not like the Salvation Army, I venture to say, is no justification for withholding your sympathy and practical co-operation in carrying out a Scheme which promises so much blessedness to your fellow-men. You may not like our government, our methods, our faith. Your feeling towards us might perhaps be duly described by an observation that slipped unwittingly from the tongue of a somewhat celebrated leader in the evangelistic world sometime ago, who, when asked what he thought of the Salvation Army, replied that " He did not like it at all, but he believed that God Almighty did." Perhaps, as an agency, we may not be exactly of your way of thinking, but that is hardly the question. Look at that dark ocean, full of human wrecks, writhing in anguish and despair. How to rescue those unfortunates is the question. The particular character of the methods employed, the peculiar uniforms worn by the life-boat crew, the noises made by the rocket apparatus, and the mingled shoutings of the rescued and the rescuers, may all be contrary to your taste and traditions. But all these objections and antipathies, I submit, are as nothing compared with the delivering of the people out of that dark sea.

If among my readers there be any who have the least conception that this scheme is put forward by me from any interested motives by all means let them refuse to contribute even by a single penny to what would be, at least, one of the most shameless of shams. There may be those who are able to imagine that men who have been literally martyred in this cause have faced their death for the sake of the paltry coppers they collected to keep body and soul together. Such may possibly find no difficulty in persuading themselves that this is but another attempt to raise money to augment that mythical fortune which I, who never yet drew a penny beyond mere out-of-pocket expenses from the Salvation Army funds, am supposed to be accumulating. From all such I ask only the tribute of their abuse, assured that the worst they say of me is too mild to describe the infamy of my conduct if they are correct in this interpretation of my motives.

There appears to me to be only two reasons that will justify any man, with a heart in his bosom, in refusing to co-operate with me in this Scheme :—

I. *That he should have an honest and intelligent conviction that it cannot be carried out with any reasonable measure of success; or,*

2. *That he (the objector) is prepared with some other plan which will as effectually accomplish the end it contemplates.*

Let me consider the second reason first. If it be that you have some plan that promises more directly to accomplish the deliverance of these multitudes than mine, I implore you at once to bring it out. Let it see the light of day. Let us not only hear your theory, but see the evidences which prove its practical character and assure its success. If your plan will bear investigation, I shall then consider you to be relieved from the obligation to assist me—nay, if after full consideration of your plan I find it better than mine, I will give up mine, turn to, and help you with all my might. But if you have nothing to offer, I demand your help in the name of those whose cause I plead.

Now, then, for your first objection, which I suppose can be expressed in one word—" impossible." This, if well founded, is equally fatal to my proposals. But, in reply, I may say—How do you know ? Have you inquired ? I will assume that you have read the book, and duly considered it. Surely you would not dismiss so important a theme without some thought. And though my arguments may not have sufficient weight to carry conviction, you must admit them to be of sufficient importance to warrant investigation. Will you therefore come and see for yourself what has been done already, or, rather, what we are doing to-day. Failing this, will you send someone capable of judging on your behalf. I do not care very much whom you send. It is true the things of the Spirit are spiritually discerned, but the things of humanity any man can judge, whether saint or sinner, if he only possess average intelligence and ordinary bowels of compassion.

I should, however, if I had a choice, prefer an investigator who has some practical knowledge of social economics, and much more should I be pleased if he had spent some of his own time and a little of his own money in trying to do the work himself. After such investigation I am confident there could be only one result.

There is one more plea I have to offer to those who might seek to excuse themselves from rendering any financial assistance to the Scheme. *Is it not worthy at least of being tried as an experiment ?* Tens of thousands of pounds are yearly spent in " trying" for minerals, boring for coals, sinking for water, and I believe there are those who think it worth while, at an expenditure of hundreds of thousands of pounds, to experiment in order to test the possibility of making a tunnel under the sea between this country and

France. Should these adventurers fail in their varied operations, they have, at least, the satisfaction of knowing, though hundreds of thousands of pounds have been expended, that they have not been wasted, and they will not complain; because they have at least attempted the accomplishment of that which they felt ought to be done; and it must be better to attempt a duty, though we fail, than never to attempt it at all. In this book we do think we have presented a sufficient reason to justify the expenditure of the money and effort involved in the making of this experiment. And though the effort should not terminate in the grand success which I so confidently predict, and which we all must so ardently desire, still there is bound to be, not only the satisfaction of having attempted some sort of deliverance for these wretched people, but certain results which will amply repay every farthing expended in the experiment.

I am now sixty-one years of age. The last eighteen months, during which the continual partner of all my activities for now nearly forty years has laid in the arms of unspeakable suffering, has added more than many many former ones, to the exhaustion of my term of service. I feel already something of the pressure which led the dying Emperor of Germany to say, "I have no time to be weary." If I am to see the accomplishment in any considerable degree of these life-long hopes, I must be enabled to embark upon the enterprise without delay, and with the world-wide burden constantly upon me in connection with the universal mission of our Army I cannot be expected to struggle in this matter alone.

But I trust that the upper and middle classes are at last being awakened out of their long slumber with regard to the permanent improvement of the lot of those who have hitherto been regarded as being for ever abandoned and hopeless. Shame indeed upon England if, with the example presented to us nowadays by the Emperor and Government of Germany, we simply shrug our shoulders, and pass on again to our business or our pleasure leaving these wretched multitudes in the gutters where they have lain so long. No, no, no; time is short. Let us arise in the name of God and humanity, and wipe away the sad stigma from the British banner that our horses are better treated than our labourers.

It will be seen that this Scheme contains many branches. It is probable that some of my readers may not be able to endorse the plan as a whole, while heartily approving of some of its features;

and to the support of what they do not heartily approve they may not be willing to subscribe. Where this is so, we shall be glad for them to assist us in carrying out those portions of the undertaking which more especially command their sympathy and commend themselves to their judgment. For instance, one man may believe in the Over-Sea Colony, but feel no interest in the Inebriates' Home ; another, who may not care for emigration, may desire to furnish a Factory or Rescue Home ; a third may wish to give us an estate, assist in the Food and Shelter work, or the extension of the Slum Brigade. Now, although I regard the Scheme as one and indivisible—from which you cannot take away any portion without impairing the prospect of the whole—it is quite practicable to administer the money subscribed so that the wishes of each donor may be carried out. Subscriptions may, therefore, be sent in for the general fund of the Social Scheme, or they can be devoted to any of the following distinct funds :—

1. The City Colony.
2. The Farm Colony.
3. The Colony Over-Sea.
4. The Household Salvage Brigade.
5. The Rescue Homes for Fallen Women.
6. Deliverance for the Drunkard.
7. The Prison Gate Brigade.
8. The Poor Man's Bank.
9. The Poor Man's Lawyer.
10. Whitechapel-by-the-Sea.

Or any other department suggested by the foregoing.

In making this appeal I have, so far, addressed myself chiefly to those who have money ; but money, indispensable as it is, has never been the thing most needful. Money is the sinews of war ; and, as society is at present constituted, neither carnal nor spiritual wars can be carried on without money. But there is something more necessary still. War cannot be waged without soldiers. A Wellington can do far more in a campaign than a Rothschild. More than money—a long, long way—I want men ; and when I say men, I mean women also—men of experience, men of brains, men of heart, and men of God.

In this great expedition, though I am starting for territory which is familiar enough, I am, in a certain sense, entering an unknown land. My people will be new at it. We have trained our soldiers to the saving of souls, we have taught them Knee-drill, we have instructed them in the art and mystery of dealing with the consciences and hearts of men ; and that will ever continue the main business of their lives.

To save the soul, to regenerate the life, and to inspire the spirit with the undying love of Christ is the work to which all other duties must ever be strictly subordinate in the Soldiers of the Salvation Army. But the new sphere on which we are entering will call for faculties other than those which have hitherto been cultivated, and for knowledge of a different character ; and those who have these gifts, and who are possessed of this practical information, will be sorely needed.

Already our world-wide Salvation work engrosses the energies of every Officer whom we command. With its extension we have the greatest difficulty to keep pace ; and, when this Scheme has to be practically grappled with, we shall be in greater straits than ever. True, it will find employment for a multitude of energies and talents which are now lying dormant, but, nevertheless, this extension will tax our resources to the very utmost. In view of this, reinforcements will be indispensable. We shall need the best brains, the largest experience, and the most undaunted energy of the community.

I want Recruits, but I cannot soften the conditions in order to attract men to the Colours. I want no comrades on these terms, but those who know our rules and are prepared to submit to our discipline: who are one with us on the great principles which determine our action, and whose hearts are in this great work for the amelioration of the hard lot of the lapsed and lost. These I will welcome to the service.

It may be that you cannot deliver an open-air address, or conduct an indoor meeting. Public labour for souls has hitherto been outside your practice. In the Lord's vineyard, however, are many labourers, and all are not needed to do the same thing. If you have a practical acquaintance with any of the varied operations of which I have spoken in this book ; if you are familiar with agriculture, understand the building trade, or have a practical knowledge of almost any form of manufacture, there is a place for you.

We cannot offer you great pay, social position, or any glitter and tinsel of man's glory; in fact, we can promise little more than rations, plenty of hard work, and probably no little of worldly scorn ; but if on the whole you believe you can in no other way help your Lord so well and bless humanity so much, you will brave the opposition of friends, abandon earthly prospects, trample pride under foot, and come out and follow Him *in this New Crusade.*

To you who believe in the remedy here proposed, and the soundness of these plans, and have the ability to assist me, I now confidently appeal for practical evidence of the faith that is in you. The responsibility is no longer mine alone. It is yours as much as mine. It is yours even more than mine if you withhold the means by which I may carry out the Scheme. I give what I have. If you give what you have the work will be done. If it is not done, and the dark river of wretchedness rolls on, as wide and deep as ever, the consequences will lie at the door of him who holds back.

I am only one man among my fellows, the same as you. The obligation to care for these lost and perishing multitudes does not rest on me any more than it does on you. To me has been given the idea, but to you the means by which it may be realised. The Plan has now been published to the world ; it is for you to say whether it is to remain barren, or whether it is to bear fruit in unnumbered blessings to all the children of men.

APPENDIX.

APPENDIX.

THE SALVATION ARMY.

THE POSITION OF OUR FORCES

OCTOBER, 1890.

	Corps or Socie.ies.	Out-posts.	Officers or Persons who'ly engaged in the Work.
The United Kingdom ...	1375	—	4506
France ... ... }	106	72	352
Switzerland ...			
Sweden	103	41	328
United States	363	57	1066
anada	317	78	1021
Australia—			
Victoria ... ... }			
South Australia			
New SouthWales }	270	465	903
Tasmania ...			
Queensland ...			
New Zealand	65	99	186
India ... ... }	80	51	419
Ceylon			
Holland	40	8	131
Denmark	33	—	87
Norway	45	7	132
Germany	16	6	75
Belgium	4	—	21
Finland	3	—	12
The Argentine Republic	2	—	15
South Africa and St.			
Helena	52	12	162
Total abroad	1499	896	4910
Grand total	2874	896	9416

THE SUPPLY ("TRADE") DEPARTMENT.

	At Home.	Abroad.
Buildings occupied—At Home, 8; Abroad, 22		
Officers	53	15
Employés	207	55
Total ...	260	70

THE PROPERTY DEPARTMENT.

Property now Vested in the Army ;—

The United Kingdom	£377,500
France and Switzerland	10,000
Sweden	13,593
Norway	11,676
The United States	6,601
Canada...	98,728
Australia	86,251
New Zealand	14,798
India	5,537
Holland	7,188
Denmark	2,340
South Africa...	10,101
Total	£611,618

Value of trade effects, stock, machinery, and goods on hand, £130,000 additional.

SOCIAL WORK OF THE ARMY.

Rescue homes (fallen women)...	33
Slum Posts	33
Prison Gate Brigades	10
Food Depôts	4
Shelters for the Destitute	5
Inebriates' Home	1
Factory for the " out of work"...	1
Lab ur Bureaux	2
Officers and others managing those branches	384

SALVATION AND SOCIAL REFORM LITERATURE.

	At home.	Abroad.	Circulation.
Weekly Newspapers ...	3	24	31,000,000
Monthly Magazines ...	3	12	2,400,000
Total	6	36	33,400,000

T

Total annual circulation of the above 33,400,000
Total annual circulation of other
publications 4,000,000

Total annual circulation of Army
literature... 37,400,000

THE UNITED KINGDOM—
"The War Cry" ... ... 300,000 weekly.
"The Young Soldier" ... 126,750 „
"All the World "... ... ... 50,000 monthly.
"The Deliverer"... ... ... 48,000 „

GENERAL STATEMENTS AND STATISTICS.

	Accom-modation.	Annual cost.
Training Garrisons for Officers (United Kingdom)...	28 400	£11,500
Do. Do. (Abroad)...	33 760	
Large Vans for Evangelising the Villages(known as Cavalry Forts)	7	
Homes of Rest for Officers	24 240	10,000
Indoor Meetings, held weekly	28,351	
Open-air Meetings held weekly (chiefly in England and Colonies)	21,467	
Total Meetings held weekly	49,818	
Number of Houses visited weekly (Great Britain only).............................	54,000	
Number of Countries and Colonies occupied		34
Number of Languages in which Literature is issued		15
Number of Languages in which Salvation is preached by the Officers		29
Number of Local (Non-Commissioned Officers) and Bandsmen		23,069
Number of Scribes and Office Employés		471

Average weekly reception of telegrams,
600, and letters, 5,400, at the London
Headquarters.
Sum raised annually from all sources by
the Army £750,000
BALANCE SHEETS, duly audited by chartered
accountants, are issued annually in connection
with the International Headquarters. See the
Annual Report of 1889—" Apostolic Warfare."
Balance Sheets are also produced quarterly at
every Corps in the world, audited and signed
by the Local Officers. Divisional Balance Sheets
issued monthly and audited by a Special Department at Headquarters.

Duly and independently audited Balance
Sheets are also issued annually from every
Territorial Headquarters.

THE AUXILIARY LEAGUE.

The Salvation Army International Auxiliary
League is composed

1.—Of persons who, without necessarily endorsing or approving of every single method
used by the Salvation Army, are sufficiently in
sympathy with its great work of reclaiming
drunkards, rescuing the fallen—in a word,
saving the lost—as to give it their PRAYERS,
INFLUENCE, AND MONEY.

2.—Of persons who, although seeing eye to
eye with the Army, yet are unable to join it,
owing to being actively engaged in the work of
their own denominations, or by reason of bad
health or other infirmities, which forbid their
taking any active part in Christian work.
Persons are enrolled either as Subscribing of
Collecting Auxiliaries.

The League comprises persons of influence and
position, members of nearly all denominations,
and many ministers.

PAMPHLETS.—Auxiliaries will always be
supplied gratis with copies of our Annual Report and Balance Sheet and other pamphlets
for distribution on application to Headquarters.
Some of our Auxiliaries have materially helped
us in this way by distributing our literature at
the seaside and elsewhere, and by making
arrangements for the regular supply of waiting
rooms, hydropathics, and hotels, thus helping
to dispel the prejudice under which many
persons unacquainted with the Army are found
to labour.

"ALL THE WORLD" is posted free regularly
each month to Auxiliaries.

For further information, and for full particulars of the work of The Salvation Army, apply
personally or by letter to GENERAL BOOTH,
or to the Financial Secretary at International
Headquarters, 101, Queen Victoria St., London,
E.C., to whom also contributions should be
sent.

Cheques and Postal Orders crossed "City
Bank."

THE SALVATION ARMY: A SKETCH.

BY AN OFFICER OF SEVENTEEN YEARS' STANDING.

What is the Salvation Army?

It is an Organisation existing to effect a radical revolution in the spiritual condition of the enormous majority of the people of all lands. Its aim is to produce a change not only in the opinions, feelings, and principles of these vast populations, but to alter the whole course of their lives, so that instead of spending their time in frivolity and pleasure-seeking, if not in the grossest forms of vice, they shall spend it in the service of their generation and in the worship of God. So far it has mainly operated in professedly Christian countries, where the overwhelming majority of the people have ceased, publicly, at any rate, to worship Jesus Christ, or to submit themselves in any way to His authority. To what extent has the Army succeeded?

Its flag is now flying in 34 countries or colonies, where, under the leadership of nearly 10,000 men and women, whose lives are entirely given up to the work, it is holding some 49,800 religious meetings every week, attended by millions of persons, who ten years ago would have laughed at the idea of praying. And these operations are but the means for further extension, as will be seen, especially when it is remembered that the Army has its 27 weekly newspapers, of which no less than 31,000,000 copies are sold in the streets, public-houses, and popular resorts of the godless majority. From its ranks it is therefore certain that an ever-increasing multitude of men and women must eventually be won.

That all this has not amounted to the creation of a mere passing gust of feeling, may best be demonstrated perhaps from the fact that the Army has accumulated no less than £775,000 worth of property, pays rentals amounting to £220,000 per annum for its meeting places, and has a total income from all sources of three-quarters of a million per annum.

Now consider from whence all this has sprung.

It is only twenty-five years since the author of this volume stood absolutely alone in the East of London, to endeavour to Christianise its irreligious multitudes, without the remotest conception in his own mind of the possibility of any such Organisation being created.

Consider, moreover, through what opposition the Salvation Army has ever had to make its way.

In each country it has to face universal prejudice, distrust, and contempt, and often stronger antipathy still. This opposition has generally found expression in systematic, Governmental, and Police restriction, followed in too many cases by imprisonment, and by the condemnatory outpourings of Bishops, Clergy, Pressmen and others, naturally followed in too many instances by the oaths and curses, the blows and insults of the populace. Through all this, in country after country, the Army makes its way to the position of universal respect, that respect, at any rate, which is shown to those who have conquered.

And of what material has this conquering host been made?

Wherever the Army goes it gathers into its meetings, in the first instance, a crowd of the most debased, brutal, blasphemous elements that can be found who, if permitted, interrupt the services, and if they see the slightest sign of police tolerance for their misconduct, frequently fall upon the Army officers or their property with violence. Yet a couple of Officers face such an audience with the absolute certainty of recruiting out of it an Army Corps. Many thousands of those who are now most prominent in the ranks of the Army never knew what it was to pray before they attended its services; and large numbers of them had settled into a profound conviction that everything connected with religion was utterly false. It is out of such material that God has constructed what is admitted to be one of the most fervid bodies of believers ever seen on the face of the earth.

Many persons in looking at the progress of the Army have shown a strange want of discernment in talking and writing as though all this had been done in a most haphazard fashion, or as though an individual could by the mere effort of his will produce such changes in the lives of others as he chose. The slightest reflection will be sufficient we are sure to convince any impartial individual that the gigantic results attained by the Salvation Army could only be reached by steady unaltering processes adapted to this end. And what are the processes by which this great Army has been made?

I. The foundation of all the Army's success, looked at apart from its divine source of strength, is its continued direct attack upon those whom it seeks to bring under the influence of the Gospel. The Salvation Army Officer, instead of standing upon some dignified pedestal, to describe the fallen condition of his fellow men, in the hope that though far from him, they may thus, by some mysterious process, come to a better life, goes down into the street, and from door to door, and from room to room, lays his hands on those who are spiritually sick, and leads them to the Almighty Healer. In its forms of speech and writing the Army constantly exhibits this same characteristic. Instead of propounding religious theories or pretending to teach a system of theology, it speaks much after the fashion of the old Prophet or Apostle, to each individual, about his or her sin and duty, thus bringing to bear upon each heart and conscience the light and power from heaven, by which alone the world can be transformed.

2. And step by step, along with this human contact goes unmistakably something that is not human.

The puzzlement and self-contradiction of most critics of the Army springs undoubtedly from the fact that they are bound to account for its success without admitting that any superhuman power attends its ministry, yet day after day, and night after night, the wonderful facts go on multiplying. The man who last night was drunk in a London slum, is to-night standing up for Christ on an Army platform. The clever sceptic, who a few weeks ago was interrupting the speakers in Berlin, and pouring contempt upon their claims to a personal knowledge of the unseen Saviour, is to-day as thorough a believer as any of them. The poor girl, lost to shame and hope, who a month ago was an outcast of Paris, is to-day a modest devoted follower of Christ, working in a humble situation. To those who admit we are right in saying "this is the Lord's doing," all is simple enough, and our certainty that the dregs of Society can become its ornaments requires no further explanation.

3. All these modern miracles would, however, have been comparatively useless but for the Army's system of utilising the gifts and energy of our converts to the uttermost. Suppose that without any claim to Divine power the Army had succeeded in raising up tens of thousands of persons, formerly unknown and unseen in the community, and made them into Singers, Speakers, Musicians, and Orderlies, that would surely in itself have been a remarkable fact. But not only have these engaged in various labours for the benefit of the community. They have been filled with a burning ambition to attain the highest possible degree of usefulness. No one can wonder that we expect to see the same process carried on successfully amongst our new friends of the Casual Ward and the Slum. And if the Army has been able to accomplish all this utilisation of human talents for the highest purposes, in spite of an almost universally prevailing contrary practice amongst the Churches, what may not its Social Wing be expected to do, with the example of the Army before it ?

4. The maintenance of all this system has, of course, been largely due to the unqualified acceptance of military government and discipline. But for this, we cannot be blind to the fact that even in our own ranks difficulties would every day arise as to the exaltation to front seats of those who were formerly persecutors and injurious. The old feeling which would have kept Paul suspected, in the background, after his conversion is, unfortunately, a part of the conservative groundwork of human nature that con-

tinues to exist everywhere, and which has to be overcome by rigid discipline in order to secure that everywhere and always, the new convert should be made the most of for Christ. But our Army system is a great indisputable fact, so much so that our enemies sometimes reproach us with it. That it should be possible to create an Army Organisation, and to secure faithful execution of duty daily is indeed a wonder, but a wonder accomplished, just as completely amongst the Republicans of America and France, as amongst the militarily trained Germans, or the subjects of the British monarchy. It is notorious that we can send an officer from London, possessed of no extraordinary ability, to take command of any corps in the world, with a certainty that he will find soldiers eager to do his bidding, and without a thought of disputing his commands, so long as he continues faithful to the orders and regulations under which his men are enlisted.

5. But those show a curious ignorance who set down our successes to this discipline, as though it were something of the prison order, although enforced without any of the power lying either behind the prison warder or the Catholic priest. On the contrary, wherever the discipline of the Army has been endangered, and its regular success for a time interrupted, it has been through an attempt to enforce it without enough of that joyous, cheerful spirit of love which is its main spring. Nobody can become acquainted with our soldiers in any land, without being almost immediately struck with their extraordinary gladness, and this joy is in itself one of the most infectious and influential elements of the Army's success. But if this be so, amid the comparatively well to do, judge of what its results are likely to be amongst the poorest and most wretched! To those who have never known bright days, the mere sight of a happy face is as it were a revelation and inspiration in one.

6. But the Army's success does not come with magical rapidity; it depends, like that of all real work, upon infinite perseverance.

To say nothing of the perseverance of the Officer who has made the saving of men his life work, and who, occupied and absorbed with this great pursuit, may naturally enough be expected to remain faithful, there are multitudes of our Soldiers who, after a hard day's toil for their daily bread, have but a few hours of leisure, but devote it ungrudgingly to the service of the War. Again and again, when the remains of some Soldier are laid to rest, amid the almost universal respect of a town, which once knew him only as an evil-doer, we hear it said that this man, since the date of his conversion, from five to ten years ago, has seldom been absent from his post, and never without good reason for it. His duty may have been comparatively insignificant, "only a door-keeper," "only a *War Cry* seller," yet Sunday after Sunday, evening after evening, he would be present, no matter who the commanding officer might be, to do his part, bearing with the unruly, breathing hope into the distressed, and showing unwavering faithfulness to all.

The continuance of these processes of mercy depends largely upon leadership, and the creation and maintenance of this leadership has been one of the marvels of the Movement. We have men to-day looked up to and reverenced over wide areas of country, arousing multitudes to the most devoted service, who a few years ago were champions of iniquity, notorious in nearly every form of vice, and some of them ringleaders in violent opposition to the Army. We have a right to believe that on the same lines God is going to raise up just such leaders without measure and without end.

Beneath, behind, and pervading all the successes of the Salvation Army is a force against which the world may sneer, but without which the world's miseries cannot be removed, the force of that Divine love which breathed on Calvary, and which God is able to communicate by His spirit to human hearts to-day.

It is pitiful to see intelligent men attempting to account, without the admission of this great fact, for the self-sacrifice and success of Salvation Officers and Soldiers. If those who wish to understand the Army would only take the trouble to spend as much as twenty-four hours with its people, how different in almost every instance would be the conclusions arrived at. Half-an-hour spent in the rooms inhabited by many of our officers would be sufficient to convince, even a well-to-do working man, that life could not be lived happily in such circumstances without some superhuman power, which alike sustains and gladdens the soul, altogether independently of earthly surroundings.

The Scheme that has been propounded in this volume would, we are quite satisfied, have no chance of success were it not for the fact that we have such a vast supply of men and women who, through the love of Christ ruling in their hearts, are prepared to look upon a life of self-sacrificing effort for the benefit of the vilest and roughest as the highest of privileges. With such a force at command, we dare to say that the accomplishment of this stupendous undertaking is a foregone conclusion, if the material assistance which the Army does not possess is forthcoming.

THE SALVATION ARMY SOCIAL REFORM WING.

Temporary Headquarters—

36, UPPER THAMES STREET, LONDON, E.C.

OBJECTS.—The bringing together of employers and workers for their mutual advantage. Making known the wants of each to each by providing a ready method of communication.

PLAN OF OPERATION.—The opening of a Central Registry Office, which for the present will be located at the above address, and where registers will be kept *free of charge* wherein the wants of both employers and workers will be recorded, the registers being open for consultation by all interested.

Public Waiting Rooms (for male and female), to which the unemployed may come for the purpose of scanning the newspapers, the insertion of advertisements for employment in all newspapers at lowest rates. Writing tables, &c., provided for their use to enable them to write applications for situations or work. The receiving of letters (replies to applications for employment) for unemployed workers.

The Waiting Rooms will also act as Houses-of-Call, where employers can meet and enter into engagements with Workers of all kinds, by appointment or otherwise, thus doing away with the snare that awaits many of the unemployed, who have no place to wait other than the Public House, which at present is almost the only " house-of-call " for Out-of-Work men.

By making known to the public generally the wants of the unemployed by means of advertisements, by circulars, and direct application to employers, the issue of labour statistics with information as to the number of unemployed who are anxious for work, the various trades and occupations they represent, &c., &c.

The opening of branches of the Labour Bureau as fast as funds and opportunities permit, in all the large towns and centres of industry throughout Great Britain.

In connection with the Labour Bureau, we propose to deal with both skilled and unskilled workers, amongst the latter forming such agencies as " Sandwich " Board Men's Society, Shoe Black, Carpet Beating, White-washing, Window

Cleaning, Wood Chopping, and other Brigades, all of which will, with many others, be put into operation as far as the assistance of the public (in the shape of applying for workers of all kinds) will afford us the opportunity.

A Domestic Servants' Agency will also be a branch of the Bureau, and a Home For Domestic Servants out of situation is also in contemplation. In this and other matters funds alone are required to commence operations.

All communications, donations, etc., should be addressed as above, marked " Labour Bureau," etc.

CENTRAL LABOUR BUREAU.

LOCAL AGENTS AND CORRESPONDENTS' DEPARTMENT.

Dear Comrade, — The enclosed letter, which has been sent to our Officers throughout the Field, will explain the object we have in view. Your name has been suggested to us as one whose heart is thoroughly in sympathy with any effort on behalf of poor suffering humanity. We are anxious to have in connection with each of our Corps, and in every locality throughout the Kingdom, some sympathetic, level-headed comrade, acting as our Agent or local Correspondent, to whom we could refer at all times for reliable information, and who would take it as work of love to regularly communicate useful information respecting the social condition of things generally in their neighbourhood.

Kindly reply, giving us your views and feelings on the subject as soon as possible, as we are anxious to organise at once. The first business on hand is for us to get information of those out of work and employers requiring workers, so that we can place them upon our registers, and make known the wants both of employers and employés.

We shall be glad of a communication from you, giving us some facts as to the condition of things in your locality, or any ideas or suggestions you would like to give, calculated to help us in connection with this good work.

I may say that the Social Wing not only comprehends the labour question, but also prison rescue and other branches of Salvation work, dealing with broken-down humanity generally, so that you can see what a great blessing you may be to the work of God by co-operating with us.

Believe me to be,
Yours affectionately for the Suffering and Lost, etc.

LOCAL AGENTS AND CORRESPONDENTS' DEPARTMENT.

PROPOSITION FOR LOCAL AGENT, CORRESPONDENT, ETC.

Name _____

Address _____

Occupation _____

If a Soldier, what Corps ? _____

If not a Soldier, what Denomination ? _____

If spoken to on the subject, what reply they have made ? _____

_____

_____

_____

_____

_____

Signed _____

Corps _____

Date _____ 189 .

Kindly return this as soon as possible, and we will then place ourselves in communication with the Comrade you propose for this position.

TO EMPLOYERS OF LABOUR.

M_____

We beg to bring to your notice the fact that the Salvation Army has opened at the above address (in connection with the Social Reform Wing), a Labour Bureau for the Registration of the wants of all classes of Labour, for both employer and employé in London and throughout the Kingdom, our object being to place in communication with each other, for mutual advantage, those who want workers and those who want work.

Arrangements have been made at the above address for waiting rooms, where employers can see unemployed men and women, and where the latter may have accommodation to write letters, see the advertisements in the papers, &c., &c.

If you are in want of workers of any kind, will you kindly fill up the enclosed form and return it to us? We will then have the particulars entered up, and endeavour to have your wants supplied. All applications, I need hardly assure you, will have our best attention, whether they refer to work of a permanent or temporary character.

We shall also be glad, through the information office of Labour Department, to give you any further information as to our plans, &c., or an Officer will wait upon you to receive instructions for the supply of workers, if requested.

As no charge will be made for registration of either the wants of employers or the wants of the unemployed, it will be obvious that a considerable outlay will be necessary to sustain these operations in active usefulness, and that therefore financial help will be greatly needed.

We shall gratefully receive donations, from the smallest coin up, to help to cover the cost of working this department. We think it right to say that only in special cases shall we feel at liberty to give personal recommendations. This, however, will no doubt be understood, seeing that we shall have to deal with very large numbers who are total strangers to us.

Please address all communications or donations as above, marked "Central Labour Bureau," etc.

WE PROPOSE TO ENTER UPON A CRUSADE AGAINST "SWEATING." WILL YOU HELP US?

Dear Sir,—In connection with the Social Reform Wing a Central Labour Bureau has been opened, one department of which will deal especially with that class of labour termed "unskilled," from amongst whom are drawn BOARD-MEN, MESSENGERS, BILL DISTRIBUTORS, CIRCULAR ADDRESSERS, WINDOW CLEANERS, WHITE-WASHERS, CARPET BEATERS, &c., &c.

It is very important that work given to these workers and others not enumerated, should be taxed as little as possible by the Contractor, or those who act between the employer and the worker.

In all our operations in this capacity we do not propose to make profit out of those we benefit; paying over the whole amount received, less say one half-penny in the shilling, or some such small sum which will go towards the expense of providing boards for "sandwich" boardmen, the hire of barrows, purchase of necessary tools, &c., &c.

We are very anxious to help that most needy class, the "boardmen," many of whom are "sweated" out of their miserable earnings; receiving often as low as *one shilling for a day's toil.*

WE APPEAL TO ALL WHO SYMPATHISE WITH SUFFERING HUMANITY, especially Religious and Philanthropic individuals and Societies, to assist us in our efforts, by placing orders for the supply of Boardmen, Messengers, Bill-distributors, Window-cleaners and other kinds of labour in our hands. Our charge for "boardmen" will be 2s. 2d., including boards, the placing and proper supervision of the men, &c. Two shillings, at least, will go direct to the men; most of the hirers of boardmen pay this, and some even more, but often not more than one-half reaches the men.

We shall be glad to forward you further information of our plans, or will send a representative to further explain, or to take orders, on receiving notice from you to that effect.

Believe me to be,

Yours faithfully, etc.

CENTRAL LABOUR BUREAU.

TO THE UNEMPLOYED.—MALE AND FEMALE.

NOTICE.

A Free Registry, for all kinds of unemployed labour, has been opened at the above address.

If you want work, call and make yourself and your wants known.

Enter your name and address and wants on the Registers, or fill up form below, and hand it in at above address.

Look over the advertising pages of the papers provided. Tables with pens and ink are provided for you to write for situations.

If you live at a distance, fill up this form giving all particulars, or references, and forward to Commissioner Smith, care of the Labour Bureau.

Name_____

Address_____

_____

Kind of work wanted_____

Wages you ask_____

Name.	
Age.	
During past 10 years have you had regular employment ?	
How long for ?	
What kind of work ?	
What work can you do ?	
What have you worked at at odd times ?	
How much did you earn when regularly employed ?	
How much did you earn when irregularly employed ?	
Are you married ?	
Is wife living ?	
How many children and ages ?	
If you were put on a Farm to work at anything you could do, and were supplied with food, lodging, and clothes, with a view to getting you on your feet, would you do all you could ?	

HOW BEGGARY WAS ABOLISHED IN BAVARIA BY COUNT RUMFORD.

Count Rumford was an American officer who served with considerable distinction in the Revolutionary War in that country, and afterwards settled in England. From thence he went to Bavaria, where he was promoted to the chief command of its army, and also was energetically employed in the Civil Government. Bavaria at this time literally swarmed with beggars, who were not only an eyesore and discredit to the nation, but a positive injury to the State. The Count resolved upon the extinction of this miserable profession, and the following extracts from his writings describe the method by which he accomplished it :—

"Bavaria, by the neglect of the Government, and the abuse of the kindness and charity of its amiable people, had become infested with beggars, with whom mingled vagabonds and thieves. They were to the body politic what parasites and vermin are to people and dwellings—breeding by the same lazy neglect."

—(Page 14.)

"In Bavaria there were laws which made provision for the poor, but they suffered them to fall into neglect. Beggary had become general."

—(Page 15.)

"In short," says Count Rumford, "these detestable vermin swarmed everywhere ; and not only their impudence and clamorous importunity were boundless, but they had recourse to the most diabolical arts and the most horrid crimes in the prosecution of their infamous trade. They exposed and tortured their own children, and those they stole for the purpose, to extort contributions from the charitable."

—(Page 15.)

"In the large towns beggary was an organised imposture, with a sort of government and police of its own. Each beggar had his beat, with orderly successions and promotions, as with other governments. There were battles to decide conflicting claims, and a good beat was not unfrequently a marriage portion or a thumping legacy."

(Page 16.)

" He saw that it was not enough to forbid beggary by law or to punish it by imprisonment. The beggars cared for neither. The energetic Yankee States-man attacked the question as he did problems in physical science. He studied beggary and beggars. How would he deal with one individual beggar ? Send him for a month to prison to beg again as soon as he came out ? That is no remedy. The evident course was to forbid him to beg, but at the same time to give him the opportunity to labor; to teach him to work, to encourage him to honest industry. And the wise ruler sets himself to provide food, comfort, and work for every beggar and vagabond in Bavaria, and did it."

—(Page 17.)

" Count Rumford, wise and just, sets himself to reform the whole class of beggars and vagabonds, and convert them into useful citizens, even those who had sunk into vice and crime.

" ' What,' he asked himself, ' is, after the necessaries of life, the first condition of comfort ?' Cleanliness, which animals and insects prize, which in man affects his moral character, and which is akin to godliness. The idea that the soul is defiled and depraved by what is unclean has long prevailed in all ages. Virtue never dwelt long with filth. Our bodies are at war with everything that defiles them.

" His first step, after a thorough study and consideration of the subject, was to provide in Munich, and at all necessary points, large, airy, and even elegant Houses of Industry, and store them with the tools and materials of such manu-factures as were most needed, and would be most useful. Each house was provided with a large dining-room and a cooking apparatus sufficient to furnish an economical dinner to every worker. Teachers were engaged for each kind of labour. Warmth, light, comfort, neatness, and order, in and around these houses, made them attractive. The dinner every day was gratis, provided at first by the Government, later by the contributions of the citizens. Bakers brought stale bread ; butchers, refuse meat ; citizens, their broken victuals—all rejoicing in being freed from the nuisance of beggary. The teachers of handi-crafts were provided by the Government. And while all this was free, every-one was paid the full value for his labour. You shall not beg; but here is com-fort, food, work, pay. There was no ill-usage, no harsh language ; in five years not a blow was given even to a child by his instructor.

" When the preparations for this great experiment had been silently completed, the army—the right arm of the governing power, which had been prepared for the work by its own thorough reformation—was called into action in aid of the police and the civil magistrates. Regiments of cavalry were so disposed as to furnish every town with a detachment, with patrols on every highway, and squads in the villages, keeping the strictest order and discipline, paying the utmost deference to the civil authorities, and avoiding all offence to the people ;

U

instructed when the order was given to arrest every beggar, vagrant, and deserter, and bring them before the magistrates. This military police cost nothing extra to the country beyond a few cantonments, and this expense to the whole country was less than £3,000 a-year.

"The 1st of January, 1790—New Year's Day, from time immemorial the beggars' holiday, when they swarmed in the streets, expecting everyone to give—the commissioned and non-commissioned officers of three regiments of infantry were distributed early in the morning at different points of Munich to wait for orders. Lieutenant-General Count Rumford assembled at his residence the chief officers of the army and principal magistrates of the city, and communicated to them his plans for the campaign. Then, dressed in the uniform of his rank, with his orders and decorations glittering on his breast, setting an example to the humblest soldier, he led them into the street, and had scarcely reached it before a beggar approached, wished him a 'Happy New Year,' and waited for the expected alms. 'I went up to him,' says Count Rumford, 'and laying my hand gently on his shoulder, told him that henceforth begging would not be permitted in Munich; that if he was in need, assistance would be given him; and if detected begging again, he would be severely punished.' He was then sent to the Town Hall, his name and residence inscribed upon the register, and he was directed to repair to the Military House of Industry next morning, where he would find dinner, work, and wages. Every officer, every magistrate, every soldier, followed the example set them; every beggar was arrested, and in one day a stop was put to beggary in Bavaria. It was banished out of the kingdom.

"And now let us see what was the progress and success of this experiment. It seemed a risk to trust the raw materials of industry—wool, flax, hemp, etc.—to the hands of common beggars; to render a debauched and depraved class orderly and useful, was an arduous enterprise. Of course the greater number made bad work at the beginning. For months they cost more than they came to. They spoiled more horns than they made spoons. Employed first in the coarser and ruder manufactures, they were advanced as they improved, and were for some time paid more than they earned—paid to encourage good will, effort, and perseverance. These were worth any sum. The poor people saw that they were treated with more than justice—with kindness. It was very evident that it was all for their good. At first there was confusion, but no insubordination. They were awkward, but not insensible to kindness. The aged, the weak, and the children were put to the easiest tasks. The younger children were paid simply to look on until they begged to join in the work, which seemed to them like play. Everything around them was made clean, quiet, orderly, and pleasant. Living at their own homes, they came at a fixed hour in the morning. They had at noon a hot, nourishing dinner of soup and bread. Provisions were either contributed or bought wholesale, and the economies of cookery were carried to the last point of perfection. Count Rumford had so planned the cooking apparatus that three women cooked a dinner for one thousand persons at a cost, though wood was used, of 4½d. for fuel; and the entire cost of the dinner for 1,200 was only £1 7s. 6½d., or about one-third of a penny for each person! Perfect order was kept—at work, at meals, and everywhere. As soon as a company took its place at table, the food having been previously served, all repeated a short prayer. 'Perhaps,' says Count Rumford, 'I ought to ask pardon for mentioning so old-fashioned a custom, but I own I am old-fashioned enough myself to like such things.'

"These poor people were generously paid for their labour, but something more than cash payment was necessary. There was needed the feeling of emulation, the desire to excel, the sense of honour, the love of glory. Not only pay, but rewards, prizes, distinctions, were given to the more deserving. Peculiar care was taken with the children. They were first paid simply for being present, idle lookers-on, until they begged with tears to be allowed to work. 'How sweet those tears were to me,' says Count Rumford, 'can easily be imagined.' Certain hours were spent by them in a school, for which teachers were provided.

"The effect of these measures was very remarkable. Awkward as the people were, they were not stupid, and learned to work with unexpected rapidity. More wonderful was the change in their manners, appearances and the very expression of their countenances. Cheerfulness and gratitude replaced the gloom of misery and the sullenness of despair. Their hearts were softened; they were most grateful to their benefactor for themselves, still more for their children. These worked with their parents, forming little industrial groups, whose affection excited the interest of every visitor. Parents were happy in the industry and growing intelligence of their children, and the children were proud of their own achievements.

"The great experiment was a complete and triumphant success. When Count Rumford wrote his account of it, it had been five years in operation; it was, financially, a paying speculation, and had not only banished beggary, but had wrought an entire change in the manners, habits, and very appearance of the most abandoned and degraded people in the kingdom."

("Count Rumford," pages 18-24.)

"Are the poor ungrateful? Count Rumford did not find them so. When, from the exhaustion of his great labours, he fell dangerously ill, these poor people whom he had rescued from lives of shame and misery, spontaneously assembled, formed a procession, and went in a body to the Cathedral to offer their united prayers for his recovery. When he was absent in Italy, and supposed to be dangerously ill in Naples, they set apart a certain time every day, after work hours, to pray for their benefactor. After an absence of fifteen months, Count Rumford returned with renewed health to Munich—a city where there was work for everyone, and not one person whose wants were not provided for. When he visited the military workhouse, the reception given him by these poor people drew tears from the eyes of all present. A few days after he entertained eighteen hundred of them in the English garden— a festival at which 30,000 of the citizens of Munich assisted."

("Count Rumford, pages 24-25.)

THE CO-OPERATIVE EXPERIMENT AT RALAHINE.

" The outrages of the 'Whitefeet,' 'Lady Clare Boys,' and 'Terry Alts' (labourers) far exceeded those of recent occurrence; yet no remedy but force was attempted, except by one Irish landlord, Mr. John Scott Vandeleur, of Ralahine, county Clare, late high sheriff of his county. Early in 1831 his family had been obliged to take flight, in charge of an armed police force, and his steward had been murdered by one of the labourers, having been chosen by lot at a meeting held to decide who should perpetrate the deed. Mr. Vandeleur came to England to seek someone who would aid him in organising the labourers into an agricultural and manufacturing association, to be conducted on co-operative principles, and he was recommended to Mr. Craig, who, at great sacrifice of his position and prospects, consented to give his services.

" No one but a man of rare zeal and courage would have attempted so apparently hopeless a task as that which Mr. Craig undertook. Both the men whom he had to manage—the Terry Alts who had murdered their master's steward—and their surroundings were as little calculated to give confidence in the success of the scheme as they well could be. The men spoke generally the Irish language, which Mr. Craig did not understand, and they looked upon him with suspicion as one sent to worm out of them the secret of the murder recently committed. He was consequently treated with coldness, and worse than that. On one occasion the outline of his grave was cut out of the pasture near his dwelling, and he carried his life in his hand. After a time, however, he won the confidence of these men, rendered savage as they had been by ill-treatment.

" The farm was let by Mr. Vandeleur at a fixed rent, to be paid in fixed quantities of farm produce, which, at the prices ruling in 1830-31, would bring in £900, which included interest on buildings, machinery, and live stock provided by Mr. Vandeleur. The rent alone was £700. As the farm consisted of 618 acres, only 268 of which were under tillage, this rent was a very high one—a fact which was acknowledged by the landlord. All profits after payment of rent and interest belonged to the members, divisible at the end of the year if desired. They started a co-operative store to supply themselves with food and clothing, and the estate was managed by a committee of the members, who paid every male and female member wages for their labour in labour notes which were exchangeable at the store for goods or cash. Intoxicating drink or tobacco were prohibited. The committee each day allotted each man his duties. The members worked the land partly as kitchen garden and fruit orchards, and partly as dairy farm, stall feeding being encouraged and root crops grown for the cattle. Pigs, poultry, &c., were reared. Wages at the time were only 8d per day for men and 5d. for women, and the members were paid at these rates. Yet, as they lived chiefly on potatoes and milk produced on the farm, which, as well as mutton and pork, were sold to them at extremely low prices, they saved money or rather notes. Their health and appearance quickly improved, so much so that, with disease raging round them, there was no case of death or serious illness among them while the experiment lasted. The single men lived together in a large building, and the families in cottages. Assisted by Mrs. Craig, the secretary carried out the most enlightened system of education for the young, those old enough being alternately employed on the farm and in the school. Sanitary arrangements were in a high state of perfection, and physical and moral training were most carefully attended to. In respect of these and other social arrangements, Mr. Craig was a man much before his time, and he has since made himself a name in connection with their application in various parts of the country.

"The 'New System,' as the Ralahine experiment was called, though at first regarded with suspicion and derision, quickly gained favour in the district, so that before long outsiders were extremely anxious to become members of the association. In January, 1832, the community consisted of fifty adults and seventeen children. The total number afterwards increased to eighty-one. Everything was prosperous, and the members of the association were not only benefited themselves, but their improvement exercised a beneficent influence upon the people in their neighbourhood. It was hoped that other landlords would imitate the excellent example of Mr. Vandeleur, especially as his experiment was one profitable to himself, as well as calculated to produce peace and contentment in disturbed Ireland. Just when these hopes were raised to their highest degree of expectancy, the happy community at Ralahine was broken up through the ruin and flight of Mr. Vandeleur, who had lost his property by gambling. Everything was sold off, and the labour notes saved by the members would have been worthless had not Mr. Craig, with noble self-sacrifice, redeemed them out of his own pocket.

"We have given but a very scanty description of the system pursued at Ralahine. The arrangements were in most respects admirable, and reflected the greatest credit upon Mr. Craig as an organiser and administrator. To his wisdom, energy, tact, and forbearance the success of his experiment was in great measure due, and it is greatly to be regretted that he was not in a position to repeat the attempt under more favourable circumstances."

("History of a Co-operative Farm.")

CARLYLE ON THE SOCIAL OBLIGATIONS OF THE NATION FORTY-FIVE YEARS AGO.

Inserted at the earnest request of a friend, who was struck by the coincidence of some ideas, similar to those of this volume, set forth so long ago, but as yet remaining unrealised, and which I had never read.

EXTRACTS FROM "PAST AND PRESENT."

"A Prime Minister, even here in England, who shall dare believe the heavenly omens, and address himself like a man and hero to the great dumb-struggling heart of England, and speak out for it, and act out for it, the God's-Justice it is writhing to get uttered and perishing for want of—yes, he too will see awaken round him, in passionate, burning, all-defiant loyalty, the heart of England, and such a 'support' as no Division-List or Parliamentary Majority was ever yet known to yield a man! Here as there, now as then, he who can and dare trust the heavenly Immensities, all earthly Localities are subject to him. We will pray for such a man and First-Lord;—yes, and far better, we will strive and incessantly make ready, each of us, to be worthy to serve and second such a First-Lord! We shall then be as good as sure of his arriving; sure of many things, let him arrive or not.

"Who can despair of Governments that passes a Soldier's Guard-house, or meets a red-coated man on the streets? That a body of men could be got together to kill other men when you bade them: this, *à priori*, does it not seem one of the impossiblest things? Yet look, behold it: in the stolidest of Do-nothing Governments, that impossibility is a thing done."

·   —(*Carlyle*, "Past and Present," page 223.)

"Strange, interesting, and yet most mournful to reflect on. Was this, then, of all the things mankind had some talent for, the one thing important to learn well, and bring to perfection; this of successfully killing one another? Truly, you have learned it well, and carried the business to a high perfection. It is incalculable what, by arranging, commanding, and regimenting you can make of men. These thousand straight-standing, firm-set individuals, who shoulder arms, who march, wheel, advance, retreat; and are, for your behoof a magazine charged with fiery death, in the most perfect condition of potential activity. Few months ago, till the persuasive sergeant came, what were they? Multiform ragged losels, runaway apprentices, starved weavers, thievish valets; an entirely broken population, fast tending towards the treadmill. But the persuasive sergeant came, by tap of drum enlisted, or formed lists of them, took heartily to drilling them; and he and you have made them this! Most potent effectual for all work whatsoever, is wise planning, firm, combining, and commanding among men.   Let no man despair of Governments who look on these two sentries at the Horse Guards and our United Service clubs.   I could conceive an Emigration Service, a Teaching Service, considerable varieties of United and Separate Services, of the due thousands strong, all effective as this Fighting Service is; all doing *their* work like it—which work, much more than fighting, is henceforth the necessity of these new ages we are got into!   Much lies among us, convulsively, nigh desperately, *struggling to be born*."

—(" Past and Present," page 224.)

" It was well, all this, we know; and yet it was not well.   Forty soldiers, I am told, will disperse the largest Spitalfields mob; forty to ten thousand, that is the proportion between drilled and undrilled.   Much there is which cannot yet be organised in this world, but somewhat also which can—somewhat also which must.   When one thinks, for example, what books are become and becoming for us, what operative Lancashires are become; what a Fourth Estate and innumerable virtualities not yet got to be actualities are become and becoming, one sees organisms enough in the dim huge future, and ' United Services ' quite other than the redcoat one; and much, even in these years, struggling to be born! "   —(" Past and Present," page 226.

" An effective ' Teaching Service,' I do consider that there must be; some education secretary, captain-general of teachers, who will actually contrive to get us *taught*.   Then again, why should there not be an ' Emigration Service,' and secretary with adjuncts, with funds, forces, idle navy ships, and ever-increasing apparatus, in fine an *effective system* of emigration, so that at length before our twenty years of respite ended, every honest willing workman who found England too strait, and the ' organisation of labour ' not yet sufficiently advanced, might find likewise a bridge built to carry him into new western lands, there to ' organise ' with more elbow room some labour for himself? There to be a real blessing, raising new corn for us, purchasing new webs and hatchets from us; leaving us at least in peace; instead of staying here to be a physical-force Chartist, unblessed and no blessing!   Is it not scandalous to consider that a Prime Minister could raise within the year, as I have seen it done, a hundred and twenty millions sterling to shoot the French; and we are stopped short for want of the hundredth part of that to keep the English living?   The bodies of the English living, and the souls of the English living, these two ' Services,' an Education Service and an Emigration Service, these with others, will have actually to be organised.

" A free bridge for emigrants!   Why, we should then be on a par with America itself, the most favoured of all lands that have no government; and we should have, besides, so many traditions and mementos of priceless things which

America has cast away. We could proceed deliberately to 'organise labour not doomed to perish unless we effected it within year and day every willing worker that proved superfluous, finding a bridge ready for him. This verily will have to be done; the time is big with this. Our little Isle is grown too narrow for us; but the world is wide enough yet for another six thousand years. England's sure markets will be among new colonies of Englishmen in all quarters of the Globe. All men trade with all men when mutually convenient, and are even bound to do it by the Maker of Men. Our friends of China, who guiltily refused to trade in these circumstances—had we not to argue with them, in cannon-shot at last, and convince them that they ought to trade? 'Hostile tariffs' will arise to shut us out, and then, again, will fall, to let us in; but the sons of England—speakers of the English language, were it nothing more—will in all times have the ineradicable predisposition to trade with England. Mycale was the *Pan-Ionian*—rendezvous of all the tribes of Ion—for old Greece; why should not London long continue the *All Saxon Home*, rendezvous of all the 'Children of the Harz-Rock,' arriving, in select samples, from the Antipodes and elsewhere, by steam and otherwise, to the 'season' here? What a future! Wide as the world, if we have the heart and heroism for it, which, by Heaven's blessing, we shall.

> "Keep not standing fixed and rooted,
> Briskly venture, briskly roam;
> Head and hand, where'er thou foot it,
> And stout heart are still at home.
> In what land the sun does visit
> Brisk are we, what e'er betide;
> To give space for wandering is it
> That the world was made so wide.

"Fourteen hundred years ago it was a considerable 'Emigration Service,' never doubt it, by much enlistment, discussion, and apparatus that we ourselves arrived in this remarkable island, and got into our present difficulties among others." —("Past and Present," pages 228-230.)

"The main substance of this immense problem of organising labour, and first of all of managing the working classes, will, it is very clear, have to be solved by those who stand practically in the middle of it, by those who themselves work and preside over work. Of all that can be enacted by any Parliament in regard to it, the germs must already lie potentially extant in those two classes who are to obey such enactment. A human chaos *in* which there is no light, you vainly attempt to irradiate by light shed *on* it; order never can arise there." —("Past and Present," pages 231-32.)

"Look around you. Your world-hosts are all in mutiny, in confusion, destitution; on the eve of fiery wreck and madness. They will not march farther for you, on the sixpence a day and supply-and-demand principle: they will not; nor ought they; nor can they. Ye shall reduce them to order; begin reducing them to order, to just subordination; noble loyalty in return for noble guidance. Their souls are driven nigh mad; let yours be sane and never saner. Not as a bewildered bewildering mob, but as a firm regimented mass, with real captains over them, will these men march any more. All human interests, combined human endeavours, and social growth in this world have, at a certain stage of their development, required organising; and work, the greatest of human interests, does not require it.

"God knows the task will be hard, but no noble task was ever easy. This task will wear away your lives and the lives of your sons and grandsons; but for what purpose, if not for tasks like this, were lives given to men? Ye shall cease to count your thousand-pound scalps; the noble of you shall cease! Nay, the very scalps, as I say, will not long be left, if you count only these. Ye shall cease wholly to be barbarous vulturous Choctaws, and become noble European nineteenth-century men. Ye shall know that Mammon, in never such gigs and flunky 'respectabilities' in not the alone God; that of himself he is but a devil and even a brute-god.

"Difficult? Yes, it will be difficult. The short-fibre cotton; that, too, was difficult. The waste-cotton shrub, long useless, disobedient as the thistle by the wayside; have ye not conquered it, made it into beautiful bandana webs, white woven shirts for men, bright tinted air garments wherein flit goddesses? Ye have shivered mountains asunder, made the hard iron pliant to you as putty; the forest-giants—marsh-jötuns—bear sheaves of golden grain; Ægir—the Sea-Demon himself stretches his back for a sleek highway to you, and on Firehorses and Windhorses ye career. Ye are most strong. Thor, red-bearded, with his blue sun-eyes, with his cheery heart and strong thunder-hammer, he and you have prevailed. Ye are most strong, ye Sons of icy North, of the far East, far marching from your rugged Eastern Wildernesses, hitherward from the gray dawn of Time! Ye are Sons of the *Jötun*-land; the land of Difficulties Conquered. Difficult? You must try this thing. Once try it with the understanding that it will and shall have to be done. Try it as you try the paltrier thing, making of money! I will bet on you once more, against all Jotüns, Tailor-gods, Double-barrelled Law-wards, and Denizens of Chaos whatsoever!"

—("Past and Present," pages 236-37.)

"A question here arises: Whether, in some ulterior, perhaps not far-distant stage of this 'Chivalry of Labour,' your Master-Worker may not find it possible, and needful, to grant his Workers permanent *interest* in his enterprise and theirs? So that it become, in practical result, what in essential act and justice it ever is, a joint enterprise; all men, from the Chief Master down to the lowest Overseer and Operative, economically as well as loyally concerned for it? Which question I do not answer. The answer, here or else far, is perhaps, Yes; and yet one knows the

difficulties.   Despotism is essential in most enterprises ; I am told they do not
to'erate ' freedom of debate on board a seventy-four.   Republican senate and
plebiscite would not answer well in cotton mills.   And yet, observe there too,
Freedom—not nomad's or ape's Freedom, but man's Freedom ; this is indis-
pensable.   We must have it, and will have it !   To reconcile Despotism with
Freedom—well, is that such a mystery ?   Do you not already know the way ?
It is to make your Despotism *just*.   Rigorous as Destiny, but just, too, as
Destiny and its Laws.   The Laws of God ; all men obey these, and have no
' Freedom ' at all but in obeying them.   The way is already known, part of the
way ; and courage and some qualities are needed for walking on it."

—(" Past and Present," pages 241-42.)

" Not a hay-game is this man's life, but a battle and a march, a warfare with
principalities and powers.   No idle promenade through fragrant orange-groves
and green flowery spaces, waited on by the choral Muses and rosy Hours.   It
is a stern pilgrimage through burning sandy solitudes, through regions of thick-
ribbed ice.   He walks among men, loves men with inexpressible soft pity, as
they cannot love him, but his soul dwells in solitude in the uttermost parts of
creation.   In green oases by the palm-tree wells he rests a space, but anon he
has to journey forward, escorted by the Terrors and the Splendours, the Arch-
demons and Archangels.   All Heaven, all Pandemonium are his escort.   The
stars keen-glancing from the Intensities send tidings to him ; the graves, silent
with their dead, from the Eternities.   Deep calls for him unto Deep."

—(" Past and Present," page 249.)

THE CATHOLIC CHURCH AND THE SOCIAL QUESTION.

The Rev. Dr. Earry read a paper at the Catholic Conference on June 30th, 1890, from which I take the following extracts as illustrative of the rising feeling on this subject in the Catholic Church.   The Rev. Dr. Barry began by defining the proletariat as those who have only one possession—their labour.   Those who have no land, and no stake in the land, no house, and no home except the few sticks of furniture they significantly call by the name, no right to employment, but at the most a right to poor relief; and who, until the last 20 years, had not even a right to be educated unless by the charity of their "betters."   The class which, without figure of speech or flights of rhetoric, is homeless, landless, propertyless in our chief cities—that I call the proletariat. Of the proletariat he declared there were hundreds of thousands growing up outside the pale of all churches.

He continued : For it is frightfully evident that Christianity has not kept pace with the population ; that it has lagged terribly behind ; that, in plain words, we have in our midst a nation of heathens to whom the ideals, the practices, and the commandments of religion are things unknown—as little realised in the miles on miles of tenement-houses, and the factories which have produced them, as though Christ had never lived or never died.   How could it be otherwise ? The great mass of men and women have never had time for religion.   You cannot expect them to work double-tides.   With hard physical labour, from morning till night in the surroundings we know and see, how much mind and leisure is left for higher things on six days of the week ?   .   .   . We must look this matter in the face.   I do not pretend to establish the proportion between different sections in which these things happen.   Still less am I willing to lay the blame on those who are houseless, landless, and propertyless.   What I say is that if the Government of a country allows millions of human beings to be thrown into such conditions of living and working as we have seen, these are the consequences that must be looked for. " A child," said the Anglican Bishop South, "has a right to be born, and not to be damned into the world."   Here have been millions of children literally "damned into the world," neither their heads nor their hands trained to anything useful, their miserable subsistence a thing to be fought and scrambled for, their homes reeking dens under the law of lease-holding which has produced outcast London and horrible Glasgow, their right to a playground and amusement curtailed to the running gutter, and their great "object-lesson" in life the drunken parents who end so often in the prison, the hospital, and the workhouse. We need not be astonished if these not only are not Christians, but have never understood why they should be. . . .

The social condition has created this domestic heathenism. Then the social condition must be changed. We stand in need of a public creed—of a social, and if you will understand the word, of a lay Christianity. This work cannot be done by the clergy, nor within the four walls of a church. The field of battle lies in the school, the home, the street, the tavern, the market, and wherever men come together. To make the people Christian they must be restored to their homes, and their homes to them.

M^CCORQUODALE & CO., LIMITED,
PRINTERS,
LONDON, NEWTON, LEEDS, AND GLASGOW.

PUBLICATIONS

OF THE

SALVATION ARMY.

BY MRS. BOOTH.

Practical Religion. Papers on "Training of Children;" "Worldly Amusements;" "Woman's Right to Preach;" "The Uses of Trial," etc. Price, 1s.; cloth, 1s. 6d.; gilt edges, 2s. 6d. Several of the above Addresses are also published separately. Price, 1d. each, or 6s. 6d. per 100, post free.

Aggressive Christianity. Containing, amongst others, Addresses on "Witnessing for Christ;" "Conditions of Successful Labour for Souls;" "Being Filled with the Spirit," etc. Price, 1s.; cloth boards, 1s. 6d.; gilt edges, 2s. 6d.

Godliness: Being a Report of several Addresses at St. James' Hall, London. CONTENTS: Saving Faith; Charity; Charity and Rebuke; Charity and Conflict; Charity and Loneliness; Conditions of Effectual Prayer; The Perfect Heart; How to Work for God with Success; Enthusiasm and Full Salvation; Repentance; Addresses on Holiness; Hindrances to Holiness. Price, 1s.; cloth, 1s. 6d; gilt edges, 2s. 6d.

Life and Death. Containing a series of Addresses, mainly to the Unconverted, on the following:—"The New Birth;" "Mercy and Judgment;" "Halting between Two Opinions;" "A True and a False Faith;" "Sowing and Reaping;" "The Prodigal Son;" "Quench not the Spirit;" "Save Thyself;" "The Day of His Wrath;" "Religious Indifference;" "Need of Atonement;" "A True and a False Peace;" "What is The Salvation Army?" Price, 1s.; cloth, 1s. 6d.; gilt edges, 2s. 6d.

Popular Christianity: Being a series of Lectures delivered in Princes Hall, Piccadilly, on the following subjects: "The Christs of the Nineteenth Century compared with the Christ of God;" "A Mock Salvation and a Real Deliverance from Sin;" "Sham Compassion and the Dying Love of Christ;" "Popular Christianity: Its Cowardly Service *versus* the Real Warfare;" "The Sham Judgment in contrast with the Great White Throne;" "Notes of Three Addresses on Household Gods;" "The Salvation Army Following Christ." 198 pages, paper covers, 1s.; cloth, bevelled boards, red edges, 2s.

Addresses to Business Gentlemen. Subjects: The Salvation Army: Its relation to the State, to the Churches, to Business Principles; its Future; Answers to the Main Points of Criticism on the so-called Secret Book. Price, paper, 6d.; cloth, 1s.

Holiness: Being an Address delivered in St. James' Hall, Piccadilly, London. Price, 1d.; 6s. 6d. per 100, post free.

Publications of the Salvation Army—*Continued.*

BY GENERAL BOOTH.

The General's Letters: Being a reprint of the General's weekly Letters in the "War Cry," together with Life-like Portrait of the Writer. Paper, 1s.; extra cloth boards, 2s.

Training of Children; or, How to make the Children into Saints and Soldiers of Jesus Christ. Price, limp cloth, 1s. 6d.; cloth boards, red edges, 2s. 6d.

Salvation Soldiery: A series of Addresses and Papers descriptive of the Characteristics of God's best Soldiers. With eight Illustrations. Price, 1s.; cloth boards, 1s. 6d.; cloth, extra gilt, 2s. 6d.

Holy Living; or, What The Salvation Army Teaches about Sanctification. Price, 1d.

Holiness Readings. By the GENERAL, Mrs. BOOTH, the CHIEF-OF-STAFF, Miss BOOTH, and others. Being extracts from the "Salvationist" and the "War Cry." 200 pages. Price, paper 1s.; cloth, 1s. 6d. Strongly recommended.

Orders and Regulations for the Field Officers of the Salvation Army. A complete compendium of instructions to Officers, with a statement of the doctrines and discipline of the Army, 576 pages, cloth, red edges, 5s.

All About the Salvation Army. A brief, succinct, and interesting *résumé* of the history, methods, and teaching of the Army, in the form of question and answer, 64 pages. Price, 1d., or 6s. 6d. per 100.

Orders and Regulations for Soldiers of the Salvation Army. 64 pages, 1d.

Doctrines of The Salvation Army. Limp cloth, 6d.

BY COMMISSIONER RAILTON.

Heathen England and the Salvation Army. (Fifth Edition.) This book contains full descriptions from life of the utterly godless condition of millions of the inhabitants of the British Islands, of the origin and history of The Salvation Army and its General, together with hundreds of examples of the value and success of the various operations which it carries on. Paper covers, 1s.; cloth boards, 1s. 6d.

Captain Ted: Being the Story of the Holy Life and Victorious Career of Captain Edward Irons, of the Salvation Army, drowned at Portsmouth, 1879. Paper covers, 6d.; cloth boards, 1s.

Twenty-One Years' Salvation Army. Filled with Thrilling Incidents of the War, and giving what has been so long desired by many friends—a Sketch of The Salvation Army Work from its commencement. Paper, 1s.; cloth boards, 1s. 6d.

Salvation Navvy: Being an account of the Life, Death, and Victories of Captain John Allen, of The Salvation Army. Paper, 1s.; cloth, 1s. 6d.

MISCELLANEOUS BOOKS.

Beneath Two Flags. The Aim, Methods of Work, History, and Progress of the Salvation Army in the United States. By MAUD B. BOOTH. Cloth, 3s. 6d.

The Salvation Soldiers' Guide : Being a Bible Chapter for the Morning and Evening of Every Day in the Year, with Fragments for Mid-day Reading. This book contains almost all those portions of Scripture which would be read as lessons in a public service. The four Gospels are harmonised, the historical books of the Old Testament condensed, and the genealogies, the Levitical law, and the portions of prophecy referring to particular heathen nations are omitted, so as to bring the book down to pocket size, in a type easily readable in the open air. 570 pages. Price, limp cloth, 6d. ; red cloth, red edges, 1s. ; superior red leather, gilt edges, gilt lettering, 2s. ; red French Morocco, circuit edges, gilt, 2s. 6d.

Question of Questions. By Caractacus. Paper, 1s. ; cloth, 1s. 6d.

Life of J. Nelson. A Stirring Narrative. Limp cloth, 8d.

Scriptural Way of Holiness. Paper, 1s.

What Doth Hinder? By ELIZABETH SWIFT BRENGLE. Being a series of character sketches from life, illustrating the different hindrances met with in the highway of Holiness, and showing how they may be overcome by the power of God. Paper, 6d. ; cloth, 1s.

Drum Taps : Being a Series of Sketches illustrative of the Army's peculiar operations for the rescue of the " Lapsed Masses." By E. S. B. Illustrated. Paper, 1s. ; cloth, 1s. 6d. ; gilt edges, 2s. 6d.

House-Top Saints : Being a collection of most interesting incidents in connection with Salvation work. Price, paper, 6d. ; cloth, 1s.

Life of Chas. G. Finney, The American Revivalist. A New and Revised Edition. Price, 1s. ; cloth boards, 2s.

All Sides of it. By EILEEN DOUGLAS. Being a number of Sketches of the Army's work, showing how from the lowest depths of sin it is possible to rise to the highest platform of Divine grace, and live for the salvation of others. Price 3d.

Life Links in the Warfare of Commissioner and Mrs. Booth-Tucker. Price, 6d. ; post free, 7d.

PORTRAITS.

The General.
Mrs Booth.
The Chief of the Staff.
Mrs. Bramwell Booth.
Marshal Ballington Booth.
Mrs. Ballington Booth.
Commissioner Booth-Clibborn.
Maréchale Booth-Clibborn.
Commissioner Booth-Tucker.
Mrs. Commissioner Booth-Tucker.

Commissioner Eva Booth.
Miss Lucy Booth.
Commissioner Railton.
Mrs. Commissioner Railton.
Commissioner Carleton.
Mrs. Commissioner Carleton.
Commissioner Howard.
Mrs. Commissioner Howard.
Commissioner Smith.
Mrs. Commissioner Smith.

Carte de Visites, 6d. Cabinets, 1s

MUSIC.

Salvation Army Music. Vol. I. For Soul-saving Services, Open-air Meetings, and the Home Circle. Cloth limp, 2s. 6d ; cloth boards, red edges, 3s. 6d.

Salvation Army Music. Vol. II. This Book contains none of the tunes that are to be found in Volume I. Limp cloth, 1s. ; cloth boards, 1s. 6d.

" Songs of Peace and War." Original Words and Music. (Over 30 never before published). By Commandant and Mrs. HERBERT BOOTH. With splendid Photograph. Price, 1s. 6d. each.

The Favourite Songs of the Singing, Speaking and Praying Brigade. Paper covers, Price, 1s.

The Musical Salvationist. Volumes I., II., III. and IV. Handsomely bound in cloth, Price, 2s. 6d. each volume.

The "Musical Salvationist." Volumes I., II. and III. 384 of the Newest and Best Songs of the Army. With the favourite songs of the Singing, Speaking and Praying Brigade, as a Supplement. Containing 398 pages. Price, 5s.

Brass Band Books. Printed on Waterproof paper and strongly bound, containing 88 of the most popular tunes in general use, both for indoor meetings and open-air work. Price, 9d. each instrumental part.

Brass Band Journals. Complete set for any instrument from 1 to 120. Price, 3s. each instrumental part.

Salvation Army Band Tutors. Tutors published for Cornet (1st and 2nd parts in one book) Eb Tenor ; Bb Baritone ; Eb Tenor Trombone (slide and valve) ; Bass Trombone, G ; Euphonium (Solo and Bb Bass) ; Eb Bombardon ; Bb and Eb Clarionets. The same size as the "Band Book," consisting of 36 pages, in limp cloth cover, Price, 9d. each instrumental part. These are guaranteed to embrace everything needful for Salvation Army bandsmen.

Publications of the Salvation Army—*Continued.*

The " War Cry." The Official Gazette of The Salvation Army consists of sixteen pages, sixty-four columns, with illustrations, and contains the latest intelligence of the progress of Salvation Army work in ALL PARTS OF THE WORLD. Every Saturday. Price, 1d.; post free, 1s. 8d. per quarter; 3s. 3d. per half year; 6s. 6d. per annum.

The "Young Soldier." The Salvation Army Children's "War Cry." Sixteen pages. Largely illustrated. Price ½d., or post free to any address, 1s. 1d. per quarter; 4s. 4d. per annum.

"All the World." A Monthly Magazine devoted to the record of Salvation Army work in all lands. Copiously illustrated. Price, 3d.; 4s. per annum post free. Vol. I., 1885; Vol. II., 1886; Vol. III., 1887; and Vol IV., 1888, 3s. 6d. each, net; Vol. V., 1889, 5s.

The "Deliverer." A Monthly Record of the Rescue Work of The Salvation Army. Price, 1d., 1s. 6d. per annum, post free. Volume I., July, 1889, to June, 1890, 2s. 6d.

The foregoing may be obtained by order of JOHN SNOW & CO., 2, *Ivy Lane, Paternoster Row, E.C.; any Bookseller, Railway Bookstall, or Newsagent. Also of the Officers of the various Salvation Army Corps, or direct from the Publishing Department, 98 and 100, Clerkenwell Road, London, E.C.*

WILL

CONTRIBUTORS TO THE SALVATION ARMY

PLEASE TAKE NOTICE THAT

ALL CONTRIBUTIONS for any of the various funds enumerated below should be addressed to The Financial Secretary, 101, Queen Victoria Street, London, E.C., or to General Booth, and the particular fund for which the subscriptions are intended should be specified ?

GENERAL MAINTENANCE FUND.—For the maintenance, general oversight, and extension of the work in all parts of the world.

INTERNATIONAL TRAINING HOME FUND.—To meet the cost of training and equipment of Officers.

THE BUILDING AUXILIARY FUND.—(Formerly Property League) provides advances to Corps to secure suitable buildings.

SICK AND WOUNDED FUND.—To provide Homes of Rest and necessary treatment for Officers who break down through disease, injury, or overwork.

FOREIGN SERVICE FUND.—To meet the expense connected with the sending out of Officers to all countries, and the opening up of fresh fields of battle.

THE RESCUE FUND.—For the salvation of fallen girls and women.

THE SLUM FUND.—To meet the expenses connected with the establishment and maintenance of the various Slum Posts in the lowest and poorest neighbourhoods of London and other cities.

THE FOOD AND SHELTER FUND.—To enable the various Depôts which have been established by the Army in different parts of the metropolis, for the purpose of supplying the homeless and unemployed poor with food and shelter at a nominal cost, to be maintained and increased.

CHEQUES AND POSTAL ORDERS should, in all cases, be made payable to William Booth, and crossed "City Bank."